Graduate Texts in Mathematics 103

Graduate Texts in Mathematics

continued after Index

Serge Lang

Complex Analysis

Second Edition

With 132 Illustrations

Springer-Verlag
New York Berlin Heidelberg Tokyo

Serge Lang
Department of Mathematics
Yale University
New Haven, CT 06520
U.S.A.

AMS Subject Classification: 30-01

Library of Congress Cataloging in Publication Data
Lang, Serge
  Complex analysis.
  (Graduate texts in mathematics; 103)
  Includes index.
  1. Functions of complex variables. 2. Mathematical
analysis. I. Title. II. Series.
QA331.L255 1985   515.9   84-21274

The first edition of this book was published by Addison-Wesley Publishing Co., Menlo Park, CA, in 1977.

Typeset by Composition House Ltd., Salisbury, England.
Printed and bound by R. R. Donnelley & Sons, Harrisonburg, Virginia.
Printed in the United States of America.

9 8 7 6 5 4 3 2 1

ISBN 0-387-96085-6 Springer-Verlag New York Berlin Heidelberg Tokyo
ISBN 3-540-96085-6 Springer-Verlag Berlin Heidelberg New York Tokyo

Foreword

The present book is meant as a text for a course on complex analysis at
the advanced undergraduate level, or first-year graduate level. Somewhat
more material has been included than can be covered at leisure in one
term, to give opportunities for the instructor to exercise his taste, and
lead the course in whatever direction strikes his fancy at the time.
A large number of routine exercises are included for the more standard
portions, and a few harder exercises of striking theoretical interest are
also included, but may be omitted in courses addressed to less advanced
students.

In some sense, I think the classical German prewar texts were the
best (Hurwitz–Courant, Knopp, Bieberbach, etc.) and I would recom-
mend to anyone to look through them. More recent texts have emphasized connections with real analysis, which is important, but at the cost
of exhibiting succinctly and clearly what is peculiar about complex analysis: the power series expansion, the uniqueness of analytic continuation,
and the calculus of residues. The systematic elementary development of
formal and convergent power series was standard fare in the German
texts, but only Cartan, in the more recent books, includes this material,
which I think is quite essential, e.g., for differential equations. I have
written a short text, exhibiting these features, making it applicable to a
wide variety of tastes.

The book essentially decomposes into two parts.

The *first part*, Chapters I through VIII, includes the basic properties
of analytic functions, essentially what cannot be left out of, say, a one-
semester course.

I have no fixed idea about the manner in which Cauchy's theorem is
to be treated. In less advanced classes, or if time is lacking, the usual

hand waving about simple closed curves and interiors is not entirely inappropriate. Perhaps better would be to state precisely the homological version and omit the formal proof. For those who want a more thorough understanding, I include the relevant material.

Artin originally had the idea of basing the homology needed for complex variables on the winding number. I have included his proof for Cauchy's theorem, extracting, however, a purely topological lemma of independent interest, not made explicit in Artin's original *Notre Dame* notes (cf. collected works) or in Ahlfor's book closely following Artin. I have also included the more recent proof by Dixon, which uses the winding number, but replaces the topological lemma by greater use of elementary properties of analytic functions which can be derived directly from the local theorem. The two aspects, homotopy and homology, both enter in an essential fashion for different applications of analytic functions, and neither is slighted at the expense of the other.

Most expositions usually include some of the global geometric properties of analytic maps at an early stage. I chose to make the preliminaries on complex functions as short as possible to get quickly into the analytic part of complex function theory: power series expansions and Cauchy's theorem. The advantages of doing this, reaching the heart of the subject rapidly, are obvious. The cost is that certain elementary global geometric considerations are thus omitted from Chapter I, for instance, to reappear later in connection with analytic isomorphisms (Conformal Mappings, Chapter VII) and potential theory (Harmonic Functions, Chapter VIII). I think it is best for the coherence of the book to have covered in one sweep the basic analytic material before dealing with these more geometric global topics. Since the proof of the general Riemann mapping theorem is somewhat more difficult than the study of the specific cases considered in Chapter VII, it has been postponed to the second part.

The *second part* of the book, Chapters IX through XIV, deals with further assorted analytic aspects of functions in many directions, which may lead to many other branches of analysis. I have emphasized the possibility of defining analytic functions by an integral involving a parameter and differentiating under the integral sign. Some classical functions are given to work out as exercises, but the gamma function is worked out in detail in the text, as a prototype. *The chapters in this part are essentially logically independent and can be covered in any order, or omitted at will.*

In particular, the chapter on analytic continuation, including the Schwarz reflection principle, and/or the proof of the Riemann mapping theorem could be done right after Chapter VII, and still achieve great coherence.

As most of this part is somewhat harder than the first part, it can easily be omitted from a course addressed to undergraduates. In the same spirit, some of the harder exercises in the first part have been starred, to make their omission easy.

In this second edition, I have rewritten many sections, and I have added some material. I have also made a number of corrections whose need was pointed out to me by several people. I thank them all.

I am much indebted to Barnet M. Weinstock for his help in correcting the proofs, and for useful suggestions.

SERGE LANG

Prerequisites

We assume that the reader has had two years of calculus, and has some acquaintance with epsilon–delta techniques. For convenience, we have recalled all the necessary lemmas we need for continuous functions on compact sets in the plane.

We use what is now standard terminology. A function

$$f : S \to T$$

is called **injective** if $x \neq y$ in S implies $f(x) \neq f(y)$. It is called **surjective** if for every z in T there exists $x \in S$ such that $f(x) = z$. If f is surjective, then we also say that f maps S **onto** T. If f is both injective and surjective then we say that f is **bijective**.

Given two functions f, g defined on a set of real numbers containing arbitrarily large numbers, and such that $g(x) \geq 0$, we write

$$f \ll g \quad \text{or} \quad f(x) \ll g(x) \quad \text{for} \quad x \to \infty$$

to mean that there exists a number $C > 0$ such that for all x sufficiently large, we have

$$|f(x)| \leq Cg(x).$$

Similarly, if the functions are defined for x near 0, we use the same symbol $\ll$ for $x \to 0$ to mean that there

$$|f(x)| \leq Cg(x)$$

for all x sufficiently small (there exists $\delta > 0$ such that if $|x| < \delta$ then $|f(x)| \leq Cg(x)$). Often this relation is also expressed by writing

$$f(x) = O(g(x)),$$

which is read: $f(x)$ is **big oh of** $g(x)$, for $x \to \infty$ or $x \to 0$ as the case may be.

We use $]a, b[$ to denote the **open** interval of numbers

$$a < x < b.$$

Similarly, $[a, b[$ denotes the half-open interval, etc.

Contents

PART ONE

BASIC THEORY

CHAPTER I

Complex Numbers and Functions

One of the advantages of dealing with the real numbers instead of the rational numbers is that certain equations which do not have any solutions in the rational numbers have a solution in real numbers. For instance, $x^2 = 2$ is such an equation. However, we also know some equations having no solution in real numbers, for instance $x^2 = -1$, or $x^2 = -2$. We define a new kind of number where such equations have solutions. The new kind of numbers will be called **complex** numbers.

I §1. Definition

The **complex numbers** are a set of objects which can be added and multiplied, the sum and product of two complex numbers being also a complex number, and satisfy the following conditions.

1. Every real number is a complex number, and if α, β are real numbers, then their sum and product as complex numbers are the same as their sum and product as real numbers.
2. There is a complex number denoted by i such that $i^2 = -1$.
3. Every complex number can be written uniquely in the form $a + bi$ where a, b are real numbers.
4. The ordinary laws of arithmetic concerning addition and multiplication are satisfied. We list these laws:

   If α, β, γ are complex numbers, then $(\alpha\beta)\gamma = \alpha(\beta\gamma)$, and

   $$(\alpha + \beta) + \gamma = \alpha + (\beta + \gamma).$$

We have $\alpha(\beta + \gamma) = \alpha\beta + \alpha\gamma$, and $(\beta + \gamma)\alpha = \beta\alpha + \gamma\alpha$.

We have $\alpha\beta = \beta\alpha$, and $\alpha + \beta = \beta + \alpha$.

If 1 is the real number one, then $1\alpha = \alpha$.

If 0 is the real number zero, then $0\alpha = 0$.

We have $\alpha + (-1)\alpha = 0$.

We shall now draw consequences of these properties. With each complex number $a + bi$, we associate the point (a, b) in the plane. Let $\alpha = a_1 + a_2 i$ and $\beta = b_1 + b_2 i$ be two complex numbers. Then

$$\alpha + \beta = a_1 + b_1 + (a_2 + b_2)i.$$

Hence addition of complex numbers is carried out "componentwise". For example, $(2 + 3i) + (-1 + 5i) = 1 + 8i$.

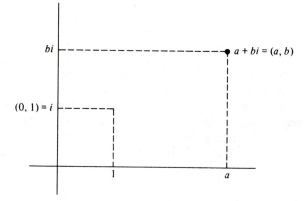

Figure 1

In multiplying complex numbers, we use the rule $i^2 = -1$ to simplify a product and to put it in the form $a + bi$. For instance, let $\alpha = 2 + 3i$ and $\beta = 1 - i$. Then

$$\begin{aligned}
\alpha\beta = (2 + 3i)(1 - i) &= 2(1 - i) + 3i(1 - i) \\
&= 2 - 2i + 3i - 3i^2 \\
&= 2 + i - 3(-1) \\
&= 2 + 3 + i \\
&= 5 + i.
\end{aligned}$$

Let $\alpha = a + bi$ be a complex number. We define $\bar{\alpha}$ to be $a - bi$. Thus if $\alpha = 2 + 3i$, then $\bar{\alpha} = 2 - 3i$. The complex number $\bar{\alpha}$ is called the **conjugate** of α. We see at once that

$$\alpha\bar{\alpha} = a^2 + b^2.$$

With the vector interpretation of complex numbers, we see that $\alpha\bar{\alpha}$ is the square of the distance of the point (a, b) from the origin.

We now have one more important property of complex numbers, which will allow us to divide by complex numbers other than 0.

If $\alpha = a + bi$ is a complex number $\neq 0$, and if we let

$$\lambda = \frac{\bar{\alpha}}{a^2 + b^2}$$

then $\alpha\lambda = \lambda\alpha = 1$.

The proof of this property is an immediate consequence of the law of multiplication of complex numbers, because

$$\alpha \frac{\bar{\alpha}}{a^2 + b^2} = \frac{\alpha\bar{\alpha}}{a^2 + b^2} = 1.$$

The number λ above is called the **inverse** of α, and is denoted by α^{-1} or $1/\alpha$. If α, β are complex numbers, we often write β/α instead of $\alpha^{-1}\beta$ (or $\beta\alpha^{-1}$), just as we did with real numbers. We see that we can divide by complex numbers $\neq 0$.

Example. To find the inverse of $(1 + i)$ we note that the conjugate of $1 + i$ is $1 - i$ and that $(1 + i)(1 - i) = 2$. Hence

$$(1 + i)^{-1} = \frac{1 - i}{2}.$$

Theorem 1.1. *Let α, β be complex numbers. Then*

$$\overline{\alpha\beta} = \bar{\alpha}\bar{\beta}, \qquad \overline{\alpha + \beta} = \bar{\alpha} + \bar{\beta}, \qquad \bar{\bar{\alpha}} = \alpha.$$

Proof. The proofs follow immediately from the definitions of addition, multiplication, and the complex conjugate. We leave them as exercises (Exercises 3 and 4).

Let $\alpha = a + bi$ be a complex number, where a, b are real. We shall call a the **real part** of α, and denote it by $\text{Re}(\alpha)$. Thus

$$\alpha + \bar{\alpha} = 2a = 2\,\text{Re}(\alpha).$$

The real number b is called the **imaginary part** of α, and denoted by $\text{Im}(\alpha)$.

We define the **absolute value** of a complex number $\alpha = a_1 + ia_2$ (where a_1, a_2 are real) to be

$$|\alpha| = \sqrt{a_1^2 + a_2^2}.$$

If we think of α as a point in the plane (a_1, a_2), then $|\alpha|$ is the length of the line segment from the origin to α. In terms of the absolute value, we can write

$$\alpha^{-1} = \frac{\bar{\alpha}}{|\alpha|^2}$$

provided $\alpha \neq 0$. Indeed, we observe that $|\alpha|^2 = \alpha\bar{\alpha}$.

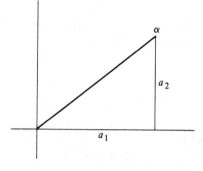

Figure 2

If $\alpha = a_1 + ia_2$, we note that

$$|\alpha| = |\bar{\alpha}|$$

because $(-a_2)^2 = a_2^2$, so $\sqrt{a_1^2 + a_2^2} = \sqrt{a_1^2 + (-a_2)^2}$.

Theorem 1.2. *The absolute value of a complex number satisfies the following properties. If α, β are complex numbers, then*

$$|\alpha\beta| = |\alpha||\beta|,$$

$$|\alpha + \beta| \leq |\alpha| + |\beta|.$$

Proof. We have

$$|\alpha\beta|^2 = \alpha\beta\overline{\alpha\beta} = \alpha\bar{\alpha}\beta\bar{\beta} = |\alpha|^2|\beta|^2.$$

Taking the square root, we conclude that $|\alpha||\beta| = |\alpha\beta|$, thus proving the first assertion. As for the second, we have

$$\begin{aligned}
|\alpha + \beta|^2 &= (\alpha + \beta)\overline{(\alpha + \beta)} = (\alpha + \beta)(\bar{\alpha} + \bar{\beta}) \\
&= \alpha\bar{\alpha} + \beta\bar{\alpha} + \alpha\bar{\beta} + \beta\bar{\beta} \\
&= |\alpha|^2 + 2\,\mathrm{Re}(\beta\bar{\alpha}) + |\beta|^2
\end{aligned}$$

because $\alpha\beta = \bar{\beta}\bar{\alpha}$. However, we have

$$2\,\mathrm{Re}(\beta\bar{\alpha}) \leq 2|\beta\bar{\alpha}|$$

because the real part of a complex number is $\leq$ its absolute value. Hence

$$\begin{aligned}|\alpha + \beta|^2 &\leq |\alpha|^2 + 2|\beta\bar{\alpha}| + |\beta|^2 \\ &\leq |\alpha|^2 + 2|\beta||\alpha| + |\beta|^2 \\ &= (|\alpha| + |\beta|)^2.\end{aligned}$$

Taking the square root yields the second assertion of the theorem.

The inequality

$$|\alpha + \beta| \leq |\alpha| + |\beta|$$

is called the **triangle inequality**. It also applies to a sum of several terms. If $z_1, \ldots, z_n$ are complex numbers then we have

$$|z_1 + \cdots + z_n| \leq |z_1| + \cdots + |z_n|.$$

Also observe that for any complex number z, we have

$$|-z| = |z|.$$

Proof?

EXERCISES I §1

1. Express the following complex numbers in the form $x + iy$, where x, y are real numbers.
   (a) $(-1 + 3i)^{-1}$
   (b) $(1 + i)(1 - i)$
   (c) $(1 + i)i(2 - i)$
   (d) $(i - 1)(2 - i)$
   (e) $(7 + \pi i)(\pi + i)$
   (f) $(2i + 1)\pi i$
   (g) $(\sqrt{2}\, i)(\pi + 3i)$
   (h) $(i + 1)(i - 2)(i + 3)$

2. Express the following complex numbers in the form $x + iy$, where x, y are real numbers.

   (a) $(1 + i)^{-1}$
   (b) $\dfrac{1}{3 + i}$
   (c) $\dfrac{2 + i}{2 - i}$
   (d) $\dfrac{1}{2 - i}$

   (e) $\dfrac{1 + i}{i}$
   (f) $\dfrac{i}{1 + i}$
   (g) $\dfrac{2i}{3 - i}$
   (h) $\dfrac{1}{-1 + i}$

3. Let α be a complex number $\neq 0$. What is the absolute value of $\alpha/\bar{\alpha}$? What is $\bar{\bar{\alpha}}$?

4. Let α, β be two complex numbers. Show that $\overline{\alpha\beta} = \bar{\alpha}\bar{\beta}$ and that

$$\overline{\alpha + \beta} = \bar{\alpha} + \bar{\beta}.$$

5. Justify the assertion made in the proof of Theorem 1.2, that the real part of a complex number is $\leq$ its absolute value.

6. If $\alpha = a + ib$ with a, b real, then b is called the **imaginary part** of α and we write $b = \text{Im}(\alpha)$. Show that $\alpha - \bar{\alpha} = 2i\,\text{Im}(\alpha)$. Show that

$$\text{Im}(\alpha) \leq |\text{Im}(\alpha)| \leq |\alpha|.$$

7. Find the real and imaginary parts of $(1 + i)^{100}$.

8. Prove that for any two complex numbers z, w we have:
   (a) $|z| \leq |z - w| + |w|$
   (b) $|z| - |w| \leq |z - w|$
   (c) $|z| - |w| \leq |z + w|$

9. Let $\alpha = a + ib$ and $z = x + iy$. Let c be real > 0. Transform the condition

$$|z - \alpha| = c$$

   into an equation involving only x, y, a, b, and c, and describe in a simple way what geometric figure is represented by this equation.

10. Describe geometrically the sets of points z satisfying the following conditions.
   (a) $|z - i + 3| = 5$ (b) $|z - i + 3| > 5$
   (c) $|z - i + 3| \leq 5$ (d) $|z + 2i| \leq 1$
   (e) $\text{Im}\, z > 0$ (f) $\text{Im}\, z \geq 0$
   (g) $\text{Re}\, z > 0$ (h) $\text{Re}\, z \geq 0$

11. Let $a_1, \ldots, a_n$ and $b_1, \ldots, b_n$ be complex numbers. Assume that $a_1, \ldots, a_n$ are distinct. Find a polynomial $P(z)$ of degree at most $n - 1$ such that $P(a_j) = b_j$ for $j = 1, \ldots, n$. Prove that such a polynomial is unique. [*Hint:* Use the Vandermonde determinant.]

I §2. Polar Form

Let $(x, y) = x + iy$ be a complex number. We know that any point in the plane can be represented by polar coordinates (r, θ). We shall now see how to write our complex number in terms of such polar coordinates.

Let θ be a real number. We define the expression $e^{i\theta}$ to be

$$e^{i\theta} = \cos\theta + i\sin\theta.$$

Thus $e^{i\theta}$ is a complex number.

For example, if $\theta = \pi$, then $e^{i\pi} = -1$. Also, $e^{2\pi i} = 1$, and $e^{i\pi/2} = i$. Furthermore, $e^{i(\theta + 2\pi)} = e^{i\theta}$ for any real θ.

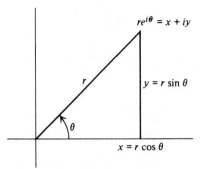

Figure 3

Let x, y be real numbers and $x + iy$ a complex number. Let

$$r = \sqrt{x^2 + y^2}.$$

If (r, θ) are the polar coordinates of the point (x, y) in the plane, then

$$x = r \cos \theta \quad \text{and} \quad y = r \sin \theta.$$

Hence

$$x + iy = r \cos \theta + ir \sin \theta = re^{i\theta}.$$

The expression $re^{i\theta}$ is called the **polar form** of the complex number $x + iy$. The number θ is sometimes called the **angle**, or **argument** of z, and we write

$$\theta = \arg z.$$

The most important property of this polar form is given in Theorem 2.1. It will allow us to have a very good geometric interpretation for the product of two complex numbers.

Theorem 2.1. *Let θ, φ be two real numbers. Then*

$$e^{i\theta + i\varphi} = e^{i\theta}e^{i\varphi}.$$

Proof. By definition, we have

$$e^{i\theta + i\varphi} = e^{i(\theta + \varphi)} = \cos(\theta + \varphi) + i \sin(\theta + \varphi).$$

Using the addition formulas for sine and cosine, we see that the preceding expression is equal to

$$\cos \theta \cos \varphi - \sin \theta \sin \varphi + i(\sin \theta \cos \varphi + \sin \varphi \cos \theta).$$

This is exactly the same expression as the one we obtain by multiplying out

$$(\cos \theta + i \sin \theta)(\cos \varphi + i \sin \varphi).$$

Our theorem is proved.

Theorem 2.1 justifies our notation, by showing that the exponential of complex numbers satisfies the same formal rule as the exponential of real numbers.

Let $\alpha = a_1 + ia_2$ be a complex number. We define e^α to be

$$e^{a_1} e^{ia_2}.$$

For instance, let $\alpha = 2 + 3i$. Then $e^\alpha = e^2 e^{3i}$.

Theorem 2.2. *Let α, β be complex numbers. Then*

$$e^{\alpha + \beta} = e^\alpha e^\beta.$$

Proof. Let $\alpha = a_1 + ia_2$ and $\beta = b_1 + ib_2$. Then

$$e^{\alpha + \beta} = e^{(a_1 + b_1) + i(a_2 + b_2)} = e^{a_1 + b_1} e^{i(a_2 + b_2)}$$
$$= e^{a_1} e^{b_1} e^{ia_2 + ib_2}.$$

Using Theorem 2.1, we see that this last expression is equal to

$$e^{a_1} e^{b_1} e^{ia_2} e^{ib_2} = e^{a_1} e^{ia_2} e^{b_1} e^{ib_2}.$$

By definition, this is equal to $e^\alpha e^\beta$, thereby proving our theorem.

Theorem 2.2 is very useful in dealing with complex numbers. We shall now consider several examples to illustrate it.

Example 1. Find a complex number whose square is $4e^{i\pi/2}$.

Let $z = 2e^{i\pi/4}$. Using the rule for exponentials, we see that $z^2 = 4e^{i\pi/2}$.

Example 2. Let n be a positive integer. Find a complex number w such that $w^n = e^{i\pi/2}$.

It is clear that complex number $w = e^{i\pi/2n}$ satisfies our requirement. In other words, we may express Theorem 2.2 as follows:

Let $z_1 = r_1 e^{i\theta_1}$ and $z_2 = r_2 e^{i\theta_2}$ be two complex numbers. To find the product $z_1 z_2$, we multiply the absolute values and add the angles. Thus

$$z_1 z_2 = r_1 r_2 e^{i(\theta_1 + \theta_2)}.$$

In many cases, this way of visualizing the product of complex numbers is more useful than that coming out of the definition.

EXERCISES I §2

1. Put the following complex numbers in polar form.

   (a) $1 + i$ (b) $1 + i\sqrt{2}$ (c) -3 (d) $4i$

   (e) $1 - i\sqrt{2}$ (f) $-5i$ (g) -7 (h) $-1 - i$

2. Put the following complex numbers in the ordinary form $x + iy$.

   (a) $e^{3i\pi}$ (b) $e^{2i\pi/3}$ (c) $3e^{i\pi/4}$ (d) $\pi e^{-i\pi/3}$

   (e) $e^{2\pi i/6}$ (f) $e^{-i\pi/2}$ (g) $e^{-i\pi}$ (h) $e^{-5i\pi/4}$

3. Let α be a complex number $\neq 0$. Show that there are two distinct complex numbers whose square is α.

4. Let $a + bi$ be a complex number. Find real numbers x, y such that

$$(x + iy)^2 = a + bi,$$

expressing x, y in terms of a and b.

5. Plot all the complex numbers z such that $z^n = 1$ on a sheet of graph paper, for $n = 2, 3, 4,$ and 5.

6. Let α be a complex number $\neq 0$. Let n be a positive integer. Show that there are n distinct complex numbers z such that $z^n = \alpha$. Write these complex numbers in polar form.

7. Find the real and imaginary parts of $i^{1/4}$, taking the fourth root such that its angle lies between 0 and $\pi/2$.

8. (a) Describe all complex numbers z such that $e^z = 1$.

   (b) Let w be a complex number. Let α be a complex number such that $e^\alpha = w$. Describe all complex numbers z such that $e^z = w$.

9. If $e^z = e^w$, show that there is an integer k such that $z = w + 2\pi ki$.

10. (a) If θ is real, show that

$$\cos\theta = \frac{e^{i\theta} + e^{-i\theta}}{2} \quad \text{and} \quad \sin\theta = \frac{e^{i\theta} - e^{-i\theta}}{2i}.$$

   (b) For arbitrary complex z, suppose we define $\cos z$ and $\sin z$ by replacing θ with z in the above formula. Show that the only values of z for which $\cos z = 0$ and $\sin z = 0$ are the usual real values from trigonometry.

11. Prove that for any complex number $z \neq 1$ we have

$$1 + z + \cdots + z^n = \frac{z^{n+1} - 1}{z - 1}.$$

12. Using the preceding exercise, and taking real parts, prove:

$$1 + \cos\theta + \cos 2\theta + \cdots + \cos n\theta = \frac{1}{2} + \frac{\sin[(n + \frac{1}{2})\theta]}{2\sin\dfrac{\theta}{2}}$$

for $0 < \theta < 2\pi$.

13. Let z, w be two complex numbers such that $\bar{z}w \neq 1$. Prove that

$$\left|\frac{z - w}{1 - \bar{z}w}\right| < 1 \quad \text{if} \quad |z| < 1 \quad \text{and} \quad |w| < 1,$$

$$\left|\frac{z - w}{1 - \bar{z}w}\right| = 1 \quad \text{if} \quad |z| = 1 \quad \text{or} \quad |w| = 1.$$

(There are many ways of doing this. One way is as follows. First check that you may assume that z is real, say $z = r$. For the first inequality you are reduced to proving

$$(r - w)(r - \bar{w}) < (1 - rw)(1 - r\bar{w}).$$

You can then use elementary calculus, differentiating with respect to r and seeing what happens for $r = 0$ and $r < 1$, to conclude the proof.)

I §3. Complex Valued Functions

Let S be a set of complex numbers. An association which to each element of S associates a complex number is called a **complex valued function**, or a **function** for short. We denote such a function by symbols like

$$f: S \to \mathbf{C}.$$

If z is an element of S, we write the association of the **value** $f(z)$ to z by the special arrow

$$z \mapsto f(z).$$

We can write

$$f(z) = u(z) + iv(z),$$

where $u(z)$ and $v(z)$ are real numbers, and thus

$$z \mapsto u(z), \qquad z \mapsto v(z)$$

are real valued functions. We call u the **real part** of f, and v the **imaginary part** of f.

We shall usually write

$$z = x + iy,$$

where x, y are real. Then the values of the function f can be written in the form

$$f(z) = f(x + iy) = u(x, y) + iv(x, y),$$

viewing u, v as functions of the two real variables x and y.

Example. For the function

$$f(z) = x^3 y + i \sin(x + y),$$

we have the real part,

$$u(x, y) = x^3 y,$$

and the imaginary part,

$$v(x, y) = \sin(x + y).$$

Example. The most important examples of complex functions are the power functions. Let n be a positive integer. Let

$$f(z) = z^n.$$

Then in polar coordinates, we can write $z = re^{i\theta}$, and therefore

$$f(z) = r^n e^{in\theta} = r^n(\cos n\theta + i \sin n\theta).$$

For this function, the real part is $r^n \cos n\theta$, and the imaginary part is $r^n \sin n\theta$.

Let D be the closed disc of radius 1 centered at the origin in **C**. In other words, D is the set of complex numbers z such that $|z| \leq 1$. If z is an element of D, then z^n is also an element of D, and so $z \mapsto z^n$ maps D into itself. Let S be the sector of complex numbers $re^{i\theta}$ such that

$$0 \leq \theta \leq 2\pi/n,$$

as shown on Fig. 4.

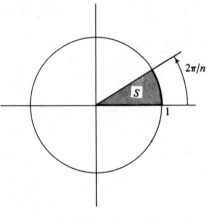

Figure 4

The function of a real variable

$$r \mapsto r^n$$

maps the unit interval $[0, 1]$ onto itself. The function

$$\theta \mapsto n\theta$$

maps the interval

$$[0, 2\pi/n] \rightarrow [0, 2\pi].$$

In this way, we see that the function $f(z) = z^n$ maps the sector S onto the full disc of all numbers

$$w = te^{i\varphi},$$

with $0 \leq t \leq 1$ and $0 \leq \varphi \leq 2\pi$. We may say that the power function wraps the sector around the disc.

   We could give a similar argument with other sectors of angle $2\pi/n$ as shown on Fig. 5. Thus we see that $z \mapsto z^n$ wraps the disc n times around.

   Given a complex number $z = re^{i\theta}$, you should have done Exercise 6 of the preceding section, or at least thought about it. For future reference, we now give the answer explicitly. We want to describe all complex numbers w such that $w^n = z$. Write

$$w = te^{i\varphi}.$$

Then

$$w^n = t^n e^{in\varphi}, \qquad 0 \leq t.$$

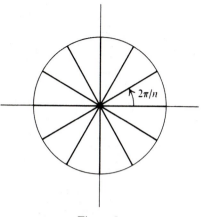

Figure 5

If $w^n = z$, then $t^n = r$, and there is a unique real number $t \geq 0$ such that $t^n = r$. On the other hand, we must also have

$$e^{in\varphi} = e^{i\theta},$$

which is equivalent with

$$in\varphi = i\theta + 2\pi ik,$$

where k is some integer. Thus we can solve for φ and get

$$\varphi = \frac{\theta}{n} + \frac{2\pi k}{n}.$$

The numbers

$$w_k = e^{i\theta/n} e^{2\pi ik/n}, \qquad k = 0, 1, \ldots, n - 1$$

are all distinct, and are drawn on Fig. 6. These numbers w_k may be described pictorially as those points on the circle which are the vertices of a regular polygon with n sides inscribed in the unit circle, with one vertex being at the point $e^{i\theta/n}$.

Each complex number

$$\zeta^k = e^{2\pi ik/n}$$

is called a **root of unity**, in fact, an n-th root of unity, because its n-th power is 1, namely

$$(\zeta^k)^n = e^{2\pi ikn/n} = e^{2\pi ik} = 1.$$

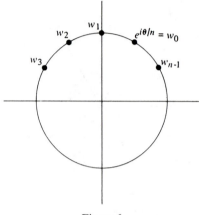

Figure 6

The points w_k are just the product of $e^{i\theta/n}$ with all the n-th roots of unity,

$$w_k = e^{i\theta/n} \zeta^k.$$

One of the major results of the theory of complex variables is to reduce the study of certain functions, including most of the common functions you can think of (like exponentials, logs, sine, cosine) to power series, which can be approximated by polynomials. Thus the power function is in some sense the unique basic function out of which the others are constructed. For this reason it was essential to get a good intuition of the power function. We postpone discussing the geometric aspects of the other functions to Chapters VII and VIII, except for some simple exercises.

EXERCISES I §3

1. Let $f(z) = 1/z$. Describe what f does to the inside and outside of the unit circle, and also what it does to points on the unit circle. This map is called **inversion** through the unit circle.

2. Let $f(z) = 1/\bar{z}$. Describe f in the same manner as in Exercise 1. This map is called **reflection** through the unit circle.

3. Let $f(z) = e^{2\pi i z}$. Describe the image under f of the set shaded in Fig. 7, consisting of those points $x + iy$ with $-\frac{1}{2} \le x \le \frac{1}{2}$ and $y \ge B$.

4. Let $f(z) = e^z$. Describe the image under f of the following sets:
   (a) The set of $z = x + iy$ such that $x \le 1$ and $0 \le y \le \pi$.
   (b) The set of $z = x + iy$ such that $0 \le y \le \pi$ (no condition on x).

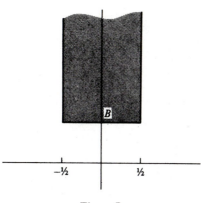

Figure 7

I §4. Limits and Compact Sets

Let α be a complex number.   By the **open disc** of radius $r > 0$ centered at α we mean the set of complex numbers z such that

$$|z - \alpha| < r.$$

For the closed disc, we use the condition $|z - \alpha| \leqq r$ instead.   We shall deal only with the open disc unless otherwise specified, and thus speak simply of the **disc**, denoted by $D(\alpha, r)$.

Let U be a subset of the complex plane. We say that U is **open** if for every point α in U there is a disc $D(\alpha, r)$ centered at α, and of some radius $r > 0$ such that this disc $D(\alpha, r)$ is contained in U.   We have illustrated an open set in Fig. 8.

Note that the radius r of the disc depends on the point α.   As α comes closer to the boundary of U, the radius of the disc will be smaller.

Examples of Open Sets.   The first quadrant, consisting of all numbers $z = x + iy$ with $x > 0$ and $y > 0$ is open, and drawn on Fig. 9.

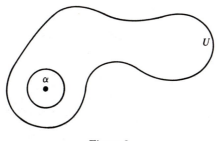

Figure 8

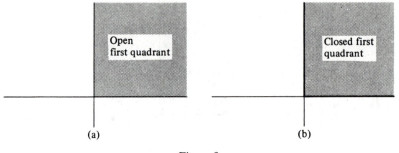

(a)   (b)

Figure 9

On the other hand, the set consisting of the first quadrant and the vertical and horizontal axes as on Fig. 9(b) is not open.

The **upper half plane** by definition is the set of complex numbers

$$z = x + iy$$

with $y > 0$. It is an open set.

Let S be a subset of the plane. A **boundary point** of S is a point α such that **every** disc $D(\alpha, r)$ centered at α and of radius $r > 0$ contains both points of S and points not in S. In the closed first quadrant of Fig. 9(b), the points on the x-axis with $x \geqq 0$ and on the y-axis with $y \geqq 0$ are boundary points of the quadrant.

A point α is said to be **adherent** to S if **every** disc $D(\alpha, r)$ with $r > 0$ contains **some** element of S. A point α is said to be an **interior** point of S if **there exists** a disc $D(\alpha, r)$ which is contained in S. Thus an adherent point can be a boundary point or an interior point of S. A set is called **closed** if it contains all its boundary points. The complement of a closed set is then open.

The **closure** of a set S is defined to be the union of S and all its boundary points. We denote the closure by S^c (and not $\bar{S}$ as it is sometimes done, in order to avoid confusions with complex conjugation).

A set S is said to be **bounded** if there exists a number $C > 0$ such that

$$|z| \leqq C \qquad \text{for all } z \text{ in } S.$$

For instance, the set in Fig. 10 is bounded. The first quadrant is not bounded.

The upper half plane is not bounded. The condition for boundedness means that the set is contained in the disc of radius C, as shown on Fig. 10.

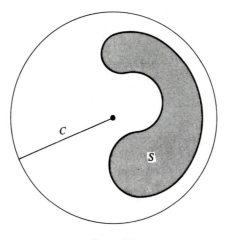

Figure 10

Let f be a function on S, and let α be an adherent point of S. Let w be a complex number. We say that

$$w = \lim_{\substack{z \to \alpha \\ z \in S}} f(z)$$

if the following condition is satisfied. Given $\epsilon > 0$ there exists $\delta > 0$ such that if $z \in S$ and $|z - \alpha| < \delta$, then

$$|f(z) - w| < \epsilon.$$

We usually omit the symbols $z \in S$ under the limit sign, and write merely

$$\lim_{z \to \alpha} f(z).$$

In some applications $\alpha \in S$ and in some applications, $\alpha \notin S$.

Let $\alpha \in S$. We say that f is **continuous** at α if

$$\lim_{z \to \alpha} f(z) = f(\alpha).$$

These definitions are completely analogous to those which you should have had in some analysis or advanced calculus course, so we don't spend much time on them. As usual, we have the rules for limits of sums, products, quotients as in calculus.

If $\{z_n\}$ $(n = 1, 2, \ldots)$ is a sequence of complex numbers, then we say that

$$w = \lim_{n \to \infty} z_n$$

if the following condition is satisfied:

Given $\epsilon > 0$ there exists an integer N such that if $n \geq N$, then

$$|z_n - w| < \epsilon.$$

Let S be the set of fractions $1/n$, with $n = 1, 2, \ldots$. Let $f(1/n) = z_n$. Then

$$\lim_{n \to \infty} z_n = w \qquad \text{if and only if} \qquad \lim_{\substack{z \to 0 \\ z \in S}} f(z) = w.$$

Thus basic properties of limits for $n \to \infty$ are reduced to similar properties for functions. Note that in this case, the number 0 is not an element of S.

A sequence $\{z_n\}$ is said to be a **Cauchy sequence** if, given ϵ, there exists N such that if $m, n \geq N$, then

$$|z_n - z_m| < \epsilon.$$

Write

$$z_n = x_n + iy_n.$$

Since

$$|z_n - z_m| = \sqrt{(x_n - x_m)^2 + (y_n - y_m)^2},$$

and

$$|x_n - x_m| \leq |z_n - z_m|, \qquad |y_n - y_m| \leq |z_n - z_m|,$$

we conclude that $\{z_n\}$ is Cauchy if and only if the sequences $\{x_n\}$ and $\{y_n\}$ of real and imaginary parts are also Cauchy. Since we know that real Cauchy sequences converge (i.e. have limits), we conclude that complex Cauchy sequences also converge.

We note that all the usual theorems about limits hold for complex numbers: Limits of sums, limits of products, limits of quotients, limits of composite functions. The proofs which you had in advanced calculus hold without change in the present context. It is then usually easy to compute limits.

Example. Find the limit

$$\lim_{n \to \infty} \frac{nz}{1 + nz}$$

for any complex number z.

If $z = 0$, it is clear that the limit is 0. Suppose $z \neq 0$. Then the quotient whose limit we are supposed to find can be written

$$\frac{nz}{1 + nz} = \frac{z}{\dfrac{1}{n} + z}.$$

But

$$\lim_{n \to \infty} \left(\frac{1}{n} + z \right) = z.$$

Hence the limit of the quotient is $z/z = 1$.

Compact Sets

We shall now go through the basic results concerning compact sets. Let S be a set of complex numbers. Let $\{z_n\}$ be a sequence in S. By a **point of accumulation of** $\{z_n\}$ we mean a complex number v such that given ϵ (always assumed > 0) there exist infinitely many integers n such that

$$|z_n - v| < \epsilon.$$

We may say that given an open set U containing v, there exist infinitely many n such that $z_n \in U$.

Similarly we define the notion of **point of accumulation** of an infinite set S. It is a complex number v such that given an open set U containing v, there exist infinitely many elements of S lying in U. In particular, a point of accumulation of S is adherent to S.

We assume that the reader is acquainted with the **Weierstrass–Bolzano theorem** about sets of real numbers: *If S is an infinite bounded set of real numbers, then S has a point of accumulation.*

We define a set of complex numbers S to be **compact** if every sequence of elements of S has a point of accumulation in S. This property is equivalent to the following properties, which could be taken as alternate definitions:

(a) Every infinite subset of S has a point of accumulation in S.
(b) Every sequence of elements of S has a convergent subsequence whose limit is in S.

We leave the proof of the equivalence between the three possible definitions to the reader.

Theorem 4.1. *A set of complex numbers is compact if and only if it is closed and bounded.*

Proof. Assume that S is compact. If S is not bounded, for each positive integer n there exists $z_n \in S$ such that

$$|z_n| > n.$$

Then the sequence $\{z_n\}$ does not have a point of accumulation. Indeed, if v is a point of accumulation, pick $m > 2|v|$, and note that $|v| > 0$. Then

$$|z_m - v| \geq |z_m| - |v| \geq m - |v| > |v|.$$

This contradicts the fact that for infinitely many m we must have z_m close to v. Hence S is bounded. To show S is closed, let v be in its closure. Given n, there exists $z_n \in S$ such that

$$|z_n - v| < 1/n.$$

The sequence $\{z_n\}$ converges to v, and has a subsequence converging to a limit in S because S is assumed compact. This limit must be v, whence $v \in S$ and S is closed.

Conversely, assume that S is closed and bounded, and let B be a bound, so $|z| \leq B$ for all $z \in S$. If we write

$$z = x + iy,$$

then $|x| \leq B$ and $|y| \leq B$. Let $\{z_n\}$ be a sequence in S, and write

$$z_n = x_n + iy_n.$$

There is a subsequence $\{z_{n_1}\}$ such that $\{x_{n_1}\}$ converges to a real number a, and there is a sub-subsequence $\{z_{n_2}\}$ such that $\{y_{n_2}\}$ converges to a real number b. Then

$$\{z_{n_2} = x_{n_2} + iy_{n_2}\}$$

converges to $a + ib$, and S is compact. This proves the theorem.

Theorem 4.2. *Let S be a compact set and let $S_1 \supset S_2 \supset \cdots$ be a sequence of non-empty closed subsets such that $S_n \supset S_{n+1}$. Then the intersection of all S_n for all $n = 1, 2, \ldots$ is not empty.*

Proof. Let $z_n \in S_n$. The sequence $\{z_n\}$ has a point of accumulation in S. Call it v. Then v is also a point of accumulation for each subsequence $\{z_k\}$ with $k \geq n$, and hence lies in the closure of S_n for each n, But S_n is assumed closed, and hence $v \in S_n$ for all n. This proves the theorem.

Theorem 4.3. *Let S be a compact set of complex numbers, and let f be a continuous function on S. Then the image of f is compact.*

Proof. Let $\{w_n\}$ be a sequence in the image of f, so that

$$w_n = f(z_n) \qquad \text{for} \qquad z_n \in S.$$

The sequence $\{z_n\}$ has a convergent subsequence $\{z_{n_k}\}$, with a limit v in S. Since f is continuous, we have

$$\lim_{k \to \infty} w_{n_k} = \lim_{k \to \infty} f(z_{n_k}) = f(v).$$

Hence the given sequence $\{w_n\}$ has a subsequence which converges in $f(S)$. This proves that $f(S)$ is compact.

Theorem 4.4. *Let S be a compact set of complex numbers, and let*

$$f : S \to \mathbf{R}$$

be a continuous function. Then f has a maximum on S, there is, there exists $v \in S$ such that $f(z) \leq f(v)$ for all $z \in S$.

Proof. By Theorem 4.3, we know that $f(S)$ is closed and bounded. Let b be its least upper bound. Then b is adherent to $f(S)$, whence in $f(S)$ because $f(S)$ is closed. So there is some $v \in S$ such that $f(v) = b$. This proves the theorem.

Remarks. In practice, one deals with a continuous function $f : S \to \mathbf{C}$ and one applies Theorem 4.4 to the absolute value of f, which is also continuous (composite of two continuous functions).

Theorem 4.5. *Let S be a compact set, and let f be a continuous function on S. Then f is uniformly continuous, i.e. given ϵ there exists δ such that whenever $z, w \in S$ and $|z - w| < \delta$, then $|f(z) - f(w)| < \epsilon$.*

Proof. Suppose the assertion of the theorem is false. Then there exists ϵ, and for each n there exists a pair of elements $z_n, w_n \in S$ such that

$$|z_n - w_n| < 1/n \qquad \text{but} \qquad |f(z_n) - f(w_n)| > \epsilon.$$

There is an infinite subset J_1 of positive integers and some $v \in S$ such that $z_n \to v$ for $n \to \infty$ and $n \in J_1$. There is an infinite subset J_2 of J_1 and $u \in S$ such that $w_n \to u$ for $n \to \infty$ and $n \in J_2$. Then, taking the limit for $n \to \infty$ and $n \in J_2$ we obtain $|u - v| = 0$ and $u = v$ because

$$|v - u| \leq |v - z_n| + |z_n - w_n| + |w_n - u|.$$

Hence $f(v) - f(u) = 0$. Furthermore,

$$|f(z_n) - f(w_n)| \leq |f(z_n) - f(v)| + |f(v) - f(u)| + |f(u) - f(w_n)|.$$

Again taking the limit as $n \to \infty$ and $n \in J_2$, we conclude that

$$f(z_n) - f(w_n)$$

approaches 0. This contradicts the assumption that

$$|f(z_n) - f(w_n)| > \epsilon,$$

and proves the theorem.

Let A, B be two sets of complex numbers. By the **distance** between them, denoted by $d(A, B)$, we mean

$$d(A, B) = \text{g.l.b.} |z - w|,$$

where the greatest lower bound g.l.b. is taken over all elements $z \in A$ and $w \in B$. If B consists of one point, we also write $d(A, w)$ instead of $d(A, B)$.

We shall leave the next two results as easy exercises.

Theorem 4.6. *Let S be a closed set of complex numbers, and let v be a complex number. There exists a point $w \in S$ such that*

$$d(S, v) = |w - v|.$$

[*Hint*: Let E be a closed disc of some suitable radius, centered at v, and consider the function $z \mapsto |z - v|$ for $z \in S \cap E$.]

Theorem 4.7. *Let K be a compact set of complex numbers, and let S be a closed set. There exist elements $z_0 \in K$ and $w_0 \in S$ such that*

$$d(K, S) = |z_0 - w_0|.$$

[*Hint*: Consider the function $z \mapsto d(S, z)$ for $z \in K$.]

Theorem 4.8. *Let S be compact. Let r be a real number >0. There exists a finite number of open discs of radius r whose union contains S.*

Proof. Suppose this is false. Let $z_1 \in S$ and let D_1 be the open disc of radius r centered at z_1. Then D_1 does not contain S, and there is some $z_2 \in S$, $z_2 \neq z_1$. Proceeding inductively, suppose we have found open discs $D_1, \ldots, D_n$ of radius r centered at points $z_1, \ldots, z_n$, respectively, such that z_{k+1} does not lie in $D_1 \cup \cdots \cup D_k$. We can then find z_{n+1} which does not lie in $D_1 \cup \cdots \cup D_n$, and we let D_{n+1} be the disc of radius r centered at z_{n+1}. Let v be a point of accumulation of the sequence $\{z_n\}$. By definition, there exist positive integers m, k with $k > m$ such that

$$|z_k - v| < r/2 \quad \text{and} \quad |z_m - v| < r/2.$$

Then $|z_k - z_m| < r$ and this contradicts the property of our sequence $\{z_n\}$ because z_k lies in the disc D_m. This proves the theorem.

Let S be a set of complex numbers, and let I be some set. Suppose that for each $i \in I$ we are given an open set U_i. We denote this association by $\{U_i\}_{i \in I}$, and call it a **family of open sets**. The **union** of the family is the set U consisting of all z such that $z \in U_i$ for some $i \in I$. We say that the family **covers** S if S is contained in this union, that is, every $z \in S$ is contained in some U_i. We then say that the family $\{U_i\}_{i \in I}$ is an **open covering** of S. If J is a subset of I, we call the family $\{U_j\}_{j \in J}$ a **subfamily**, and if it covers S also, we call it a **subcovering** of S. In particular, if

$$U_{i_1}, \ldots, U_{i_n}$$

is a finite number of the open sets U_i, we say that it is a **finite subcovering** of S if S is contained in the finite union

$$U_{i_1} \cup \cdots \cup U_{i_n}.$$

Theorem 4.9. *Let S be a compact set, and let $\{U_i\}_{i \in I}$ be an open covering of S. Then there exists a finite subcovering, that is, a finite number of open sets $U_{i_1}, \ldots, U_{i_n}$ whose union covers S.*

Proof. By Theorem 4.8, for each n there exists a finite number of open discs of radius $1/n$ which cover S. Suppose that there is no finite subcovering of S by open sets U_i. Then for each n there exists one of the open discs D_n from the preceding finite number such that $D_n \cap S$ is not covered by any finite number of open sets U_i. Let $z_n \in D_n \cap S$, and let w be a point of accumulation of the sequence $\{z_n\}$. For some index i_0 we have $w \in U_{i_0}$. By definition, U_{i_0} contains an open disc D of radius $r > 0$

centered at w. Let N be so large that $2/N < r$. There exists $n > N$ such that

$$|z_n - w| \leqq 1/N.$$

Any point of D_n is then at a distance $\leqq 2/N$ from w, and hence D_n is contained in D, and thus contained in U_{i_0}. This contradicts the hypothesis made on D_n, and proves the theorem.

EXERCISES I §4

1. Let α be a complex number of absolute value < 1. What is $\lim_{n \to \infty} \alpha^n$? Proof?

2. If $|\alpha| > 1$, does $\lim_{n \to \infty} \alpha^n$ exist? Why?

3. Show that for any complex number $z \neq 1$, we have

$$1 + z + \cdots + z^n = \frac{z^{n+1} - 1}{z - 1}.$$

If $|z| < 1$, show that

$$\lim_{n \to \infty} (1 + z + \cdots + z^n) = \frac{1}{1 - z}.$$

4. Let f be the function defined by

$$f(z) = \lim_{n \to \infty} \frac{1}{1 + n^2 z}.$$

Show that f is the characteristic function of the set $\{0\}$, that is, $f(0) = 1$, and $f(z) = 0$ if $z \neq 0$.

5. For $|z| \neq 1$ show that the following limit exists:

$$f(z) = \lim_{n \to \infty} \left(\frac{z^n - 1}{z^n + 1} \right).$$

It is possible to define $f(z)$ when $|z| = 1$ in such a way to make f continuous?

6. Let

$$f(z) = \lim_{n \to \infty} \frac{z^n}{1 + z^n}.$$

(a) What is the domain of definition of f, that is, for which complex numbers z does the limit exist?

(b) Give explicitly the values of $f(z)$ for the various z in the domain of f.

7. For $z \neq -1$, define $S(z) = \dfrac{z + 2}{z + 1}$. Let $z_1 = i$, and define

$$z_{n+1} = S(z_n).$$

Prove that $\lim z_n$ exists and equals $\sqrt{2}$. [*Hint*: Look at

$$T(w) = \frac{w - \sqrt{2}}{w + \sqrt{2}},$$

where $w = S(z)$, and proceed inductively.]

8. Let $S(z) = (z + i)/(z + 1)$. Let $z_1 = 1$ and $z_n = S(z_{n-1})$, for $n \geq 2$. Prove that $\lim z_n$ exists and equals

$$\frac{1}{\sqrt{2}}(1 + i).$$

[*Hint*: Look at

$$T(w) = \frac{w - \sqrt{i}}{w + \sqrt{i}},$$

where $w = S(z)$ and proceed inductively.]

9. There is a system to the preceding two exercises. Suppose that a, b, c, d are complex numbers with $ad - bc \neq 0$. Define a function S by

$$S(z) = \frac{az + b}{cz + d}$$

for $z \neq -d/c$.

(a) Prove that there exists a complex number z_0 such that

$$S(z_0) = z_0.$$

Such a number is called a **fixed point** of S. How many fixed points are there usually? In Exercise 7, we have $z_0 = \sqrt{2}$, and in Exercise 8, we have $z_0 = \sqrt{i}$. Check this explicitly.

(b) Let z_0, z_1 be two fixed points of S. Define

$$T(w) = \frac{w - z_0}{w - z_1}.$$

Prove that there exists a complex number λ such that

$$T(S(z)) = \lambda T(z).$$

What is λ in Exercises 7 and 8? Give λ in general in terms of a, b, c, d, z_0, z_1.

(c) Let z_1 be given arbitrarily. Discuss in the general case when you can expect a limit for the sequence $\{z_n\}$ such that

$$z_{n+1} = S(z_n).$$

The above exercises gives some basic algebraic facts about the function S, which is called a **fractional linear transformation.** We shall study such transformations again later from a more geometrical point of view.

10. Find the fixed points of the following functions:

(a) $f(z) = \dfrac{z - 3}{z + 1}$

(b) $f(z) = \dfrac{z - 4}{z + 2}$

(c) $f(z) = \dfrac{z - i}{z + 1}$

(d) $f(z) = \dfrac{2z - 3}{z + 1}$

I §5. Complex Differentiability

In studying differentiable functions of a real variable, we took such functions defined on intervals. For complex variables, we have to select domains of definition in an analogous manner.

Let U be an open set, and let z be a point of U. Let f be a function on U. We say that f is **complex differentiable** at z if the limit

$$\lim_{h \to 0} \frac{f(z + h) - f(z)}{h}$$

exists. This limit is denoted by $f'(z)$ or df/dz.

In order not to confuse the above notion of complex differentiability with real differentiability, it is also customary to call a (complex) differentiable function **holomorphic.** In this section, differentiable will always mean complex differentiable in the above sense.

The usual proofs of a first course in calculus concerning basic properties of differentiability are valid for complex differentiability. We shall run through them again.

We note that if f is differentiable at z then f is continuous at z because

$$\lim_{h \to 0} (f(z + h) - f(z)) = \lim_{h \to 0} \frac{f(z + h) - f(z)}{h} \cdot h$$

and since the limit of a product is the product of the limits, the limit on the right-hand side is equal to 0.

We let f, g be functions defined on the open set U. We assume that f, g are differentiable at z.

Sum. *The sum $f + g$ is differentiable at z, and*

$$(f + g)'(z) = f'(z) + g'(z).$$

Proof. This is immediate from the theorem that the limit of a sum is the sum of the limits.

Product. *The product fg is differentiable at z, and*

$$(fg)'(z) = f'(z)g(z) + f(z)g'(z).$$

Proof. To determine the limit of the Newton quotient

$$\frac{f(z + h)g(z + h) - f(z)g(z)}{h}$$

we write the numerator in the form

$$f(z + h)g(z + h) - f(z)g(z + h) + f(z)g(z + h) - f(z)g(z).$$

Then the Newton quotient is equal to a sum

$$\frac{f(z + h) - f(z)}{h} g(z + h) + f(z) \frac{g(z + h) - g(z)}{h}.$$

Taking the limits yields the formula.

Quotient. *If $g(z) \neq 0$, then the quotient f/g is differentiable at z, and*

$$(f/g)'(z) = \frac{g(z)f'(z) - f(z)g'(z)}{g(z)^2}.$$

Proof. This is again proved as in ordinary calculus. We first prove the differentiability of the quotient function $1/g$. We have

$$\frac{\dfrac{1}{g(z + h)} - \dfrac{1}{g(z)}}{h} = -\frac{g(z + h) - g(z)}{h} \frac{1}{g(z + h)g(z)}$$

Taking the limit yields

$$-\frac{1}{g(z)^2} g'(z),$$

which is the usual value. The general formula for a quotient is obtained from this by writing

$$f/g = f \cdot 1/g,$$

and using the rules for the derivative of a product, and the derivative of $1/g$.

Examples. As in ordinary calculus, from the formula for a product and induction, we see that for any positive integer n,

$$\frac{dz^n}{dz} = nz^{n-1}.$$

The rule for a quotient also shows that this formula remains valid when n is a negative integer.
    The derivative of $z^2/(2z - 1)$ is

$$\frac{(2z - 1)2z - 2z^2}{(2z - 1)^2}.$$

This formula is valid for any complex number z such that $2z - 1 \neq 0$. More generally, let

$$f(z) = P(z)/Q(z),$$

where P, Q are polynomials. Then f is differentiable at any point z where $Q(z) \neq 0$.
    Last comes the chain rule. Let U, V be open sets in $\mathbf{C}$, and let

$$f: U \to V \qquad \text{and} \qquad g: V \to \mathbf{C}$$

be functions, such that the image of f is contained in V. Then we can form the composite function $g \circ f$ such that

$$(g \circ f)(z) = g(f(z)).$$

Chain Rule. *Let $w = f(z)$. Assume that f is differentiable at z, and g is differentiable at w. Then $g \circ f$ is differentiable at z, and*

$$(g \circ f)'(z) = g'(f(z))f'(z).$$

Proof. Again the proof is the same as in calculus, and depends on expressing differentiability by an equivalent property not involving denominators, as follows.

Suppose that f is differentiable at z, and let

$$\varphi(h) = \frac{f(z + h) - f(z)}{h} - f'(z).$$

Then

(1) $$f(z + h) - f(z) = f'(z)h + h\varphi(h),$$

and

(2) $$\lim_{h \to 0} \varphi(h) = 0.$$

Furthermore, even though φ is at first defined only for sufficiently small h and $h \neq 0$, we may also define $\varphi(0) = 0$, and formula (1) remains valid for $h = 0$.

Conversely, suppose that there exists a function φ defined for sufficiently small h and a number a such that

(1') $$f(z + h) - f(z) = ah + h\varphi(h)$$

and

(2) $$\lim_{h \to 0} \varphi(h) = 0.$$

Dividing by h in formula (1') and taking the limit as $h \to 0$, we see that the limit exists and is equal to a. Thus $f'(z)$ exists and is equal to a. Hence the existence of a function φ satisfying (1'), (2) is equivalent to differentiability.

We apply this to a proof of the chain rule. Let $w = f(z)$, and

$$k = f(z + h) - f(z),$$

so that

$$g\left(f(z + h)\right) - g(f(z)) = g(w + k) - g(w).$$

There exists a function $\psi(k)$ such that $\lim_{k \to 0} \psi(k) = 0$ and

$$g(w + k) - g(w) = g'(w)k + k\psi(k)$$
$$= g'(w)(f(z + h) - f(z)) + (f(z + h) - f(z))\psi(k).$$

Dividing by h yields

$$\frac{g \circ f(z + h) - g \circ f(z)}{h} = g'(w)\frac{f(z + h) - f(z)}{h} + \frac{f(z + h) - f(z)}{h}\psi(k).$$

As $h \to 0$, we note that $k \to 0$ also by the continuity of f, whence $\psi(k) \to 0$ by assumption. Taking the limit of this last expression as $h \to 0$ proves the chain rule.

A function f defined on an open set U is said to be **differentiable** if it is differentiable at every point. We then also say that f is **holomorphic** on U. The word holomorphic is usually used in order not to have to specify *complex* differentiability as distinguished from real differentiability.

In line with general terminology, a holomorphic function

$$f : U \to V$$

from an open set into another is called a **holomorphic isomorphism** if there exists a holomorphic function

$$g : V \to U$$

such that g is the inverse of f, that is,

$$g \circ f = \text{id}_U \quad \text{and} \quad f \circ g = \text{id}_V.$$

A holomorphic isomorphism of U with itself is called a holomorphic **automorphism**. In the next chapter we discuss this notion in connection with functions defined by power series.

I §6. The Cauchy–Riemann Equations

Let f be a function on an open set U, and write f in terms of its real and imaginary parts,

$$f(x + iy) = u(x, y) + iv(x, y).$$

It is reasonable to ask what the condition of differentiability means in terms of u and v. We shall analyze this situation in detail in Chapter VIII, but both for the sake of tradition, and because there is some need psychologically to see right away what the answer is, we derive the equivalent conditions on u, v for f to be holomorphic.

At a fixed z, let $f'(z) = a + bi$. Let $w = h + ik$, with h, k real. Suppose that

$$f(z + w) - f(z) = f'(z)w + \sigma(w)w,$$

where

$$\lim_{w \to 0} \sigma(w) = 0.$$

Then

$$f'(z)w = (a + bi)(h + ki) = ah - bk + i(bh + ak).$$

On the other hand, let

$$F: U \to \mathbf{R}^2$$

be the map (often called vector field) such that

$$F(x, y) = (u(x, y), v(x, y)).$$

We call F the (real) **vector field associated with** f. Then

$$F(x + h, y + k) - F(x, y) = (ah - bk, bh + ak) + \sigma_1(h, k)h + \sigma_2(h, k)k,$$

where $\sigma_1(h, k)$, $\sigma_2(h, k)$ are functions tending to 0 as (h, k) tends to 0. Hence if we assume that f is holomorphic, we conclude that F is differentiable in the sense of real variables, and that its derivative is represented by the (Jacobian) matrix

$$J_F(x, y) = \begin{pmatrix} a & -b \\ b & a \end{pmatrix} = \begin{pmatrix} \dfrac{\partial u}{\partial x} & \dfrac{\partial u}{\partial y} \\ \dfrac{\partial v}{\partial x} & \dfrac{\partial v}{\partial y} \end{pmatrix}.$$

This shows that

$$f'(z) = \frac{\partial u}{\partial x} - i\frac{\partial u}{\partial y}$$

and

$$\frac{\partial u}{\partial x} = \frac{\partial v}{\partial y} \quad \text{and} \quad \frac{\partial u}{\partial y} = \frac{-\partial v}{\partial x}.$$

These are called the **Cauchy–Riemann equations**.

Conversely, let $u(x, y)$ and $v(x, y)$ be two functions satisfying the Cauchy–Riemann equations, and continuously differentiable in the sense of real functions. Define

$$f(z) = f(x + iy) = u(x, y) + iv(x, y).$$

Then it is immediately verified by reversing the above steps that f is complex-differentiable, i.e. holomorphic.

The Jacobian determinant Δ_F of the associated vector field F is

$$\Delta_F = a^2 + b^2 = \left(\frac{\partial u}{\partial x}\right)^2 + \left(\frac{\partial v}{\partial x}\right)^2 = \left(\frac{\partial u}{\partial x}\right)^2 + \left(\frac{\partial u}{\partial y}\right)^2.$$

Hence $\Delta_F \geq 0$, and is $\neq 0$ if and only if $f'(z) \neq 0$. We have

$$\boxed{\Delta_F(x, y) = |f'(z)|^2.}$$

We now drop these considerations until Chapter VIII.

The study of the real part of a holomorphic function and its relation with the function itself will be carried out more substantially in Chapter VIII. It is important, and much of that chapter depends only on elementary facts. However, the most important part of complex analysis at the present level lies in the power series aspects and the immediate applications of Cauchy's theorem. The real part plays no role in these matters. Thus we do not wish to interrupt the straightforward flow of the book now towards these topics.

However, the reader may read immediately the more elementary parts §1 and §2 of Chapter VIII, which can be understood already at this point.

EXERCISE I §6

1. Prove in detail that if u, v satisfy the Cauchy–Riemann equations, then the function

$$f(z) = f(x + iy) = u(x, y) + iv(x, y)$$

is holomorphic.

I §7. Angles Under Holomorphic Maps

In this section, we give a simple geometric property of holomorphic maps. Roughly speaking, they preserve angles. We make this precise as follows.

Let U be an open set in C and let

$$\gamma : [a, b] \to U$$

be a curve in U, so we write

$$\gamma(t) = x(t) + iy(t).$$

We assume that γ is differentiable, so its derivative is given by

$$\gamma'(t) = x'(t) + iy'(t).$$

Let $f: U \to \mathbf{C}$ be holomorphic. We let the reader verify the chain rule

$$\frac{d}{dt}(f(\gamma(t))) = f'(\gamma(t))\gamma'(t).$$

We interpret $\gamma'(t)$ as a vector in the direction of a tangent vector at the point $\gamma(t)$. This derivative $\gamma'(t)$, if not 0, defines the direction of the curve at the point.

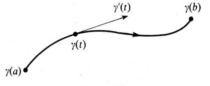

Figure 11

Consider two curves γ and η passing through the same point z_0. Say

$$z_0 = \gamma(t_0) = \eta(t_1).$$

Then the tangent vectors $\gamma'(t_0)$ and $\eta'(t_1)$ determine an angle θ which is defined to be the **angle between the curves**.

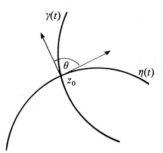

Figure 12

Applying f, the curves $f \circ \gamma$ and $f \circ \eta$ pass through the point $f(z_0)$, and by the chain rule, tangent vectors of these image curves are

$$f'(z_0)\gamma'(t_0) \quad \text{and} \quad f'(z_0)\eta'(t_1).$$

Theorem 7.1. *If $f'(z_0) \neq 0$ then the angle between the curves γ, η at z_0 is the same as the angle between the curves $f \circ \gamma, f \circ \eta$ at $f(z_0)$.*

Proof. Geometrically speaking, the tangent vectors under f are changed by multiplication with $f'(z_0)$, which can be represented in polar coordinates as a dilation and a rotation, so preserves the angles.

We shall now give a more formal argument, dealing with the cosine and sine of angles.

Let z, w be complex numbers,

$$z = a + bi \quad \text{and} \quad w = c + di,$$

where a, b, c, d are real. Then

$$z\bar{w} = ac + bd + i(bc - ad).$$

Define the **scalar product**

$$\langle z, w \rangle = \operatorname{Re}(z\bar{w}).$$

Then $\langle z, w \rangle$ is the ordinary scalar product of the vectors (a, b) and (c, d) in $\mathbf{R}^2$. Let $\theta(z, w)$ be the angle between z and w. Then

$$\cos \theta(z, w) = \frac{\langle z, w \rangle}{|z||w|}.$$

Since $\sin \theta = \cos \left(\theta - \frac{\pi}{2} \right)$, we can define

$$\sin \theta(z, w) = \frac{\langle z, -iw \rangle}{|z||w|}.$$

This gives us the desired precise formulas for the cosine and sine of an angle, which determine the angle.

Let $f'(z_0) = \alpha$. Then

$$\langle \alpha z, \alpha w \rangle = \operatorname{Re}(\alpha z \bar{\alpha} \bar{w}) = \alpha \bar{\alpha} \operatorname{Re}(z\bar{w}) = |\alpha|^2 \langle z, w \rangle$$

because $\alpha \bar{\alpha} = |\alpha|^2$ is real. It follows immediately from the above formulas that

$$\cos \theta(\alpha z, \alpha w) = \cos \theta(z, w) \quad \text{and} \quad \sin \theta(\alpha z, \alpha w) = \sin \theta(z, w).$$

This proves the theorem.

A map which preserves angles is called **conformal**. Thus we can say that a holomorphic map with non-zero derivative is conformal. The complex conjugate of a holomorphic map also preserves angles, if we disregard the orientation of an angle.

In Chapter VII, we shall consider holomorphic maps which have inverse holomorphic maps, and therefore such that their derivatives are never equal to 0. The theorem proved in this section gives additional geometric information concerning the nature of such maps. But the emphasis of the theorem in this section is local, whereas the emphasis in Chapter VII will be global. The word "conformal", however, has become a code word for this kind of map, even in the global case, which explains the title of Chapter VII. The reader will notice that the local property of preserving angles is irrelevant for the global arguments given in Chapter VII, having to do with inverse mappings. Thus in Chapter VII, we shall use a terminology which emphasizes the invertibility, namely the terminology of isomorphisms and automorphisms.

In this terminology, we can say that a holomorphic isomorphism is conformal. The converse is false in general. For instance, let U be the open set obtained by deleting the origin from the complex numbers. The function

$$f: U \to U \qquad \text{given by} \qquad z \mapsto z^2$$

has everywhere non-zero derivative in U, but it does not admit an inverse function. This function f is definitely conformal. The invertibility is true locally, however. See Theorem 1.5 of Chapter VI.

CHAPTER II

Power Series

So far, we have given only rational functions as examples of holomorphic functions. We shall study other ways of defining such functions. One of the principal ways will be by means of power series. Thus we shall see that the series

$$1 + z + \frac{z^2}{2!} + \frac{z^3}{3!} + \cdots$$

converges for all z to define a function which is equal to e^z. Similarly, we shall extend the values of $\sin z$ and $\cos z$ by their usual series to complex valued functions of a complex variable, and we shall see that they have similar properties to the functions of a real variable which you already know.

First we shall learn to manipulate power series formally. In elementary calculus courses, we derived Taylor's formula with an error term. Here we are concerned with the full power series. In a way, we pick up where calculus left off. We study systematically sums, products, inverses, and composition of power series, and then relate the formal operations with questions of convergence.

II §1. Formal Power Series

We select at first a neutral letter, say T. In writing a formal power series

$$\sum_{n=0}^{\infty} a_n T^n = a_0 + a_1 T + a_2 T^2 + \cdots$$

what is essential are its "coefficients" $a_0, a_1, a_2, \ldots$ which we shall take as complex numbers. Thus the above series may be defined as the function

$$n \longmapsto a_n$$

from the integers ≥ 0 to the complex numbers.

We could use other letters besides T, writing

$$f(T) = \sum a_n T^n,$$

$$f(r) = \sum a_n r^n,$$

$$f(z) = \sum a_n z^n.$$

In such notation, f does not denote a function, but a formal expression. Also, as a matter of notation, we write a single term

$$a_n T^n$$

to denote the power series such that $a_k = 0$ if $k \neq n$. For instance, we would write

$$5T^3$$

for the power series

$$0 + 0T + 0T^2 + 5T^3 + 0T^4 + \cdots,$$

such that $a_3 = 5$ and $a_k = 0$ if $k \neq 3$.

By definition, we call a_0 the **constant term** of f.

If

$$f = \sum a_n T^n \qquad \text{and} \qquad g = \sum b_n T^n$$

are such formal power series, we define their **sum** to be

$$f + g = \sum c_n T^n, \qquad \text{where} \quad c_n = a_n + b_n.$$

We define their **product** to be

$$fg = \sum d_n T^n, \qquad \text{where} \quad d_n = \sum_{k=0}^{n} a_k b_{n-k}.$$

The sum and product are therefore defined just as for polynomials. The first few terms of the product can be written as

$$fg = a_0 b_0 + (a_0 b_1 + a_1 b_0)T + (a_0 b_2 + a_1 b_1 + a_2 b_0)T^2 + \cdots.$$

If α is a complex number, we define

$$\alpha f = \sum (\alpha a_n) T^n$$

to be the power series whose n-th coefficient is αa_n. Thus we can multiply power series by numbers.

Just as for polynomials, one verifies that the sum and product are associative, commutative, and distributive. Thus in particular, if f, g, h are power series, then

$$f(g + h) = fg + fh \quad \text{(distributivity)}.$$

We omit the proof, which is just elementary algebra.

The **zero power series** is the series such that $a_n = 0$ for all integers $n \geq 0$.

Suppose a power series is of the form

$$f = a_r T^r + a_{r+1} T^{r+1} + \cdots,$$

and $a_r \neq 0$. Thus r is the smallest integer n such that $a_n \neq 0$. Then we call r the **order** of f, and write

$$r = \text{ord } f.$$

If ord $g = s$, so that

$$g = b_s T^s + b_{s+1} T^{s+1} + \cdots,$$

and $b_s \neq 0$, then by definition,

$$fg = a_r b_s T^{r+s} + \text{higher terms},$$

and $a_r b_s \neq 0$. Hence

$$\boxed{\text{ord } fg = \text{ord } f + \text{ord } g.}$$

A power series has order 0 if and only if it starts with a non-zero constant term. For instance, the geometric series

$$1 + T + T^2 + T^3 + \cdots$$

has order 0.

Let $f = \sum a_n T^n$ be a power series. We say that $g = \sum b_n T^n$ is an **inverse** for f if

$$fg = 1.$$

In view of the relation for orders which we just mentioned, we note that if an inverse exists, then we must have

$$\operatorname{ord} f = \operatorname{ord} g = 0.$$

In other words, both f and g start with non-zero constant terms. The converse is true:

If f has a non-zero constant term, then f has an inverse as a power series

Proof. Considering $a_0^{-1} f$ instead of f, we are reduced to the case when the constant term is equal to 1. We first note that the old geometric series gives us a formal inverse,

$$\frac{1}{1 - r} = 1 + r + r^2 + \cdots.$$

Written multiplicatively, this amounts to

$$(1 - r)(1 + r + r^2 + \cdots) = 1 + r + r^2 + \cdots - r(1 + r + r^2 + \cdots)$$
$$= 1 + r + r^2 + \cdots - r - r^2 - \cdots$$
$$= 1.$$

Next, write

$$f = 1 - h, \qquad \text{where} \quad h = -(a_1 T + a_2 T^2 + \cdots).$$

To find the inverse $(1 - h)^{-1}$ is now easy, namely it is the power series

$$(*) \qquad\qquad \varphi = 1 + h + h^2 + h^3 + \cdots.$$

We have to verify that this makes sense. Any finite sum

$$1 + h + h^2 + \cdots + h^m$$

makes sense because we have defined sums and products of power series. Observe that the order of h^n is at least n, because h^n is of the form

$$(-1)^n a_1^n T^n + \text{higher terms}.$$

Thus in the above sum (∗), if $m > n$, then the term h^m has all coefficients of order $\leq n$ equal to 0. Thus we may define the n-th coefficient of φ to be the n-th coefficient of the finite sum

$$1 + h + h^2 + \cdots + h^n.$$

It is then easy to verify that

$$(1 - h)\varphi = (1 - h)(1 + h + h^2 + h^3 + \cdots)$$

is equal to

$$1 + \text{a power series of arbitrarily high order,}$$

and consequently is equal to 1. Hence we have found the desired inverse for f.

Example. Let

$$\cos T = 1 - \frac{T^2}{2!} + \frac{T^4}{4!} - \cdots$$

be the formal power series whose coefficients are the same as for the Taylor expansion of the ordinary cosine function in elementary calculus. We want to write down the first few terms of its (formal) inverse,

$$\frac{1}{\cos T}.$$

Up to terms of order 4, these will be the same as

$$\frac{1}{1 - \left(\dfrac{T^2}{2!} - \dfrac{T^4}{4!} + \cdots\right)} = 1 + \left(\frac{T^2}{2!} - \frac{T^4}{4!} + \cdots\right)$$

$$+ \left(\frac{T^2}{2!} - \frac{T^4}{4!} + \cdots\right)^2 + \cdots$$

$$= 1 + \frac{T^2}{2!} - \frac{T^4}{4!} + \cdots + \frac{T^4}{(2!)^2} + \cdots$$

$$= 1 + \frac{1}{2}T^2 + \left(\frac{-1}{24} + \frac{1}{4}\right)T^4 + \text{higher terms.}$$

This gives us the coefficients of $1/\cos T$ up to order 4.

The substitution of h in the geometric series used to find an inverse can be generalized. Let

$$f = \sum a_n T^n$$

be a power series, and let

$$h(T) = c_1 T + \cdots$$

be a power series *whose constant term is* 0, so ord $h \geq 1$. Then we may "substitute" h in f to define a power series $f \circ h$ or $f(h)$, by

$$(f \circ h)(T) = f(h(T)) = f \circ h = a_0 + a_1 h + a_2 h^2 + a_3 h^3 + \cdots$$

in a natural way. Indeed, the finite sums

$$a_0 + a_1 h + \cdots + a_n h^n$$

are defined by the ordinary sum and product of power series. If $m > n$, then $a_m h^m$ has order $> n$; in other words, it is a power series starting with non-zero terms of order $> n$. Consequently we can define the power series $f \circ h$ as that series whose n-th coefficient is the n-th coefficient of

$$a_0 + a_1 h + \cdots + a_n h^n.$$

This composition of power series, like addition and multiplication, can therefore be computed by working only with polynomials. In fact, it is useful to discuss this approximation by polynomials a little more systematically.

We say that two power series $f = \sum a_n T^n$ and $g = \sum b_n T^n$ are **congruent mod** T^N and write $f \equiv g \pmod{T^N}$ if

$$a_n = b_n \qquad \text{for} \quad n = 0, \ldots, N - 1.$$

This means that the terms of order $\leq N - 1$ coincide for the two power series. Given the power series f, there is a unique polynomial $P(T)$ of degree $\leq N - 1$ such that

$$f(T) \equiv P(T) \pmod{T^N},$$

namely the polynomial

$$P(T) = a_0 + a_1 T + \cdots + a_{N-1} T^{N-1}.$$

If $f_1 \equiv f_2$ and $g_1 \equiv g_2$ (mod T^N), then

$$f_1 + g_1 \equiv f_2 + g_2 \quad and \quad f_1 g_1 \equiv f_2 g_2 \quad (mod\ T^N).$$

If h_1, h_2 are power series with zero constant term, and

$$h_1 \equiv h_2 \quad (mod\ T^N),$$

then

$$f_1(h_1(T)) \equiv f_2(h_2(T)) \quad (mod\ T^N).$$

Proof. We leave the sum and product to the reader. Let us look at the proof for the composition $f_1 \circ h_1$ and $f_2 \circ h_2$. First suppose h has zero constant term. Let P_1, P_2 be the polynomials of degree $N - 1$ such that

$$f_1 \equiv P_1 \quad and \quad f_2 \equiv P_2 \quad (mod\ T^N).$$

Then by hypothesis, $P_1 = P_2 = P$ is the same polynomial, and

$$f_1(h) \equiv P_1(h) = P_2(h) \equiv f_2(h) \quad (mod\ T^N).$$

Next let Q be the polynomial of degree $N - 1$ such that

$$h_1(T) \equiv h_2(T) \equiv Q(T) \quad (mod\ T^N).$$

Write $P = a_0 + a_1 T + \cdots + a_{N-1} T^{N-1}$. Then

$$P(h_1) = a_0 + a_1 h_1 + \cdots + a_{N-1} h_1^{N-1}$$

$$\equiv a_0 + a_1 Q + \cdots + a_{N-1} Q^{N-1}$$

$$\equiv a_0 + a_1 h_2 + \cdots + a_{N-1} h_2^{N-1}$$

$$\equiv P(h_2) \quad (mod\ T^N).$$

This proves the desired property, that $f_1 \circ h_1 \equiv f_2 \circ h_2$ (mod T^N).

With these rules we can compute the coefficients of various operations between power series by reducing the computations to polynomial operations, which amount to high-school algebra. Indeed, two power series f, g are equal if and only if

$$f \equiv g \quad (mod\ T^N)$$

for every positive integer N. Verifying that $f \equiv g \pmod{T^N}$ can be done by working entirely with polynomials of degree $< N$.

If f_1, f_2 are power series, then

$$(f_1 + f_2)(h) = f_1(h) + f_2(h),$$
$$(f_1 f_2)(h) = f_1(h) f_2(h).$$

If g, h have constant terms equal to 0, then

$$f(g(h)) = (f \circ g)(h).$$

Proof. In each case, the proof is obtained by reducing the statement to the polynomial case, and seeing that the required properties hold for polynomials, which is standard. For instance, for the associativity of composition, given a positive integer N, let P, Q, R be polynomials of degree $\leq N - 1$ such that

$$f \equiv P, \qquad g \equiv Q, \qquad h \equiv R \pmod{T^N}.$$

The ordinary theory of polynomials shows that

$$P(Q(R)) = (P \circ Q)(R).$$

The left-hand side is congruent to $f(g(h))$, and the right-hand side is congruent to $(f \circ g)(h) \pmod{T^N}$ by the properties which have already been proved. Hence

$$f(g(h)) \equiv (f \circ g)(h) \pmod{T^N}.$$

This is true for each N, whence $f(g(h)) = (f \circ g)(h)$, as desired.

In applications it is useful to consider power series which have a finite number of terms involving $1/z$, and this amounts also to considering arbitrary quotients of power series as follows.

Just as fractions m/n are formed with integers m, n and $n \neq 0$, we can form quotients

$$f/g = f(T)/g(T)$$

of power series such that $g \neq 0$. Two such quotients f/g and f_1/g_1 are regarded as equal if and only if

$$f g_1 = f_1 g,$$

which is exactly the condition under which we regard two rational numbers m/n and m_1/n_1 as equal. We have then defined for power series all the operations of arithmetic.

Let

$$f(T) = a_m T^m + a_{m+1} T^{m+1} + \cdots = \sum_{n \geq m} a_n T^n$$

be a power series with $a_m \neq 0$. We may then write f in the form

$$f = a_m T^m (1 + h(T)),$$

where $h(T)$ has zero constant term. Consequently $1/f$ has the form

$$1/f = \frac{1}{a_m T^m} \frac{1}{1 + h(T)}.$$

We know how to invert $1 + h(T)$, say

$$(1 + h(T))^{-1} = 1 + b_1 T + b_2 T^2 + \cdots.$$

Then $1/f(T)$ has the shape

$$1/f = a_m^{-1} \frac{1}{T^m} + a_m^{-1} b_1 \frac{1}{T^{m-1}} + \cdots.$$

It is a power series with a finite number of terms having negative powers of T. In this manner, one sees that an arbitrary quotient can always be expressed as a power series of the form

$$f/g = \frac{c_{-m}}{T^m} + \frac{c_{-m+1}}{T^{m-1}} + \cdots + c_0 + c_1 T + c_2 T^2 + \cdots$$

$$= \sum_{n \geq -m} c_n T^n.$$

If $c_{-m} \neq 0$, then we call $-m$ the **order** of f/g. It is again verified as for power series without negative terms that if

$$\varphi = f/g \quad \text{and} \quad \varphi_1 = f_1/g_1,$$

then

$$\operatorname{ord} \varphi \varphi_1 = \operatorname{ord} \varphi + \operatorname{ord} \varphi_1.$$

Example. Find the terms of order ≤ 3 in the power series for $1/\sin T$. By definition,

$$\sin T = T - T^3/3! + T^5/5! - \cdots$$
$$= T(1 - T^2/3! + T^4/5! - \cdots).$$

Hence

$$\frac{1}{\sin T} = \frac{1}{T} \frac{1}{1 - T^2/3! + T^4/5! + \cdots}$$

$$= \frac{1}{T}(1 + T^2/3! - T^4/5! + (T^2/3!\,^2 + \text{higher terms})$$

$$= \frac{1}{T} + \frac{1}{3!} T + \left(\frac{1}{(3!)^2} - \frac{1}{5!}\right) T^3 + \text{higher terms}.$$

This does what we wanted.

EXERCISES II §1

We shall write the formal power series in terms of z because that's the way they arise in practice. The series for $\sin z$, $\cos z$, e^z, etc. are to be viewed as formal series.

1. Give the terms of order ≤ 3 in the power series:

(a) $e^z \sin z$ (b) $(\sin z)(\cos z)$ (c) $\dfrac{e^z - 1}{z}$

(d) $\dfrac{e^z - \cos z}{z}$ (e) $\dfrac{1}{\cos z}$ (f) $\dfrac{\cos z}{\sin z}$

(g) $\dfrac{\sin z}{\cos z}$ (h) $e^z/\sin z$

2. Define the **Bernoulli numbers** B_n by the power series

$$\frac{z}{e^z - 1} = \sum_{n=0}^{\infty} \frac{B_n}{n!} z^n.$$

Prove the recursion formula

$$\frac{B_0}{n!\,0!} + \frac{B_1}{(n-1)!\,1!} + \cdots + \frac{B_{n-1}}{1!\,(n-1)!} = \begin{cases} 1 & \text{if } n = 1, \\ 0 & \text{if } n > 1. \end{cases}$$

Then $B_0 = 1$. Compute B_1, B_2, B_3, B_4, B_6, B_8, B_{10}, B_{12}, B_{14}. Show that $B_n = 0$ if n is odd $\neq 1$.

3. Show that

$$\frac{z}{2} \frac{e^{z/2} + e^{-z/2}}{e^{z/2} - e^{-z/2}} = \sum_{n=0}^{\infty} \frac{B_{2n}}{(2n)!} z^{2n}.$$

Replace z by $2\pi i z$ to show that

$$\pi z \cot \pi z = \sum_{n=0}^{\infty} (-1)^n \frac{(2\pi)^{2n}}{(2n)!} B_{2n} z^{2n}.$$

4. Express the power series for $\tan z$, $z/\sin z$, $z \cot z$, in terms of Bernoulli numbers.

5. **Difference Equations.** Given complex numbers a_0, a_1, a, b define a_n for $n \geq 2$ by

$$a_n = a a_{n-1} + b a_{n-2}.$$

If we have a factorization

$$T^2 - aT - b = (T - \alpha)(T - \alpha'), \quad \text{and} \quad \alpha \neq \alpha',$$

show that the numbers a_n are given by

$$a_n = A\alpha^n + B\alpha'^n$$

with suitable A, B. Find A, B in terms of a_0, a_1, α, α'. [*Hint*: Consider the power series

$$F(T) = \sum_{n=0}^{\infty} a_n T^n.$$

Show that it represents a rational function, and give its partial fraction decomposition.]

6. More generally, let $a_0, \ldots, a_{r-1}$ be given complex numbers. Let $c_0, \ldots, c_{r-1}$ be complex numbers such that the polynomial

$$P(T) = T^r - (c_{r-1} T^{r-1} + \cdots + c_0)$$

has distinct roots $\alpha_1, \ldots, \alpha_r$. Define a_n for $n \geq r$ by

$$a_n = c_0 a_{n-1} + \cdots + c_{r-1} a_{n-r}.$$

Show that there exist numbers $A_1, \ldots, A_r$ such that for all n,

$$a_n = A_1 \alpha_1^n + \cdots + A_r \alpha_r^n.$$

II §2. Convergent Power Series

We first recall some terminology about series of complex numbers.
Let $\{z_n\}$ be a sequence of complex numbers. Consider the series

$$\sum_{n=1}^{\infty} z_n.$$

We define the **partial sum**

$$s_n = \sum_{k=1}^{n} z_k = z_1 + z_2 + \cdots + z_n.$$

We say that the series **converges** if there is some w such that

$$\lim_{n \to \infty} s_n = w$$

exists, in which case we say that w is equal to the **sum of the series**,
that is,

$$w = \sum_{n=1}^{\infty} z_n.$$

If $A = \sum \alpha_n$ and $B = \sum \beta_n$ are two convergent series, with partial sums

$$s_n = \sum_{k=1}^{n} \alpha_k \quad \text{and} \quad t_n = \sum_{k=1}^{n} \beta_k,$$

then the sum and product converge. Namely,

$$A + B = \sum(\alpha_n + \beta_n)$$

$$AB = \lim_{n \to \infty} s_n t_n$$

Let $\{c_n\}$ be a series of real numbers $c_n \geq 0$. If the partial sums

$$\sum_{k=1}^{n} c_k$$

are bounded, we recall from calculus that the series converges, and that
the least upper bound of these partial sums is the limit.
Let $\sum \alpha_n$ be a series of complex numbers. We shall say that this
series **converges absolutely** if the real positive series

$$\sum |\alpha_n|$$

converges. If a series converges absolutely, then it converges. Indeed, let

$$S_n = \sum_{k=1}^{n} \alpha_k$$

be the partial sums. Then for $m \leq n$ we have

$$S_n - S_m = \alpha_{m+1} + \cdots + \alpha_n$$

whence

$$|S_n - S_m| \leq \sum_{k=m+1}^{n} |\alpha_k|.$$

Assuming absolute convergence, given ϵ there exists N such that if $n, m \geq N$, then the right-hand side of this last expression is $< \epsilon$, thereby proving that the partial sums form a Cauchy sequence, and hence that the series converges.

We have the usual test for convergence:

Let $\sum c_n$ be a series of real numbers ≥ 0 which converges. If $|\alpha_n| \leq c_n$ for all n, then the series $\sum \alpha_n$ converges absolutely.

Proof. The partial sums

$$\sum_{k=1}^{n} c_k$$

are bounded by assumption, whence the partial sums

$$\sum_{k=1}^{n} |\alpha_k| \leq \sum_{k=1}^{n} c_k$$

are also bounded, and the absolute convergence follows.

In the sequel we shall also assume some standard facts about absolutely convergent series, namely:

(i) *If a series $\sum \alpha_n$ is absolutely convergent, then the series obtained by any rearrangement of the terms is also absolutely convergent, and converges to the same limit.*

(ii) *If a double series*

$$\sum_{n=1}^{\infty} \left(\sum_{m=1}^{\infty} \alpha_{mn} \right)$$

is absolutely convergent, then the order of summation can be inter-changed, and the series so obtained is absolutely convergent, and converges to the same value.

The proof is easily obtained by considering approximating partial sums (finite sums), and estimating the tail ends. We omit it.

We shall now consider series of functions, and deal with questions of uniformity.

Let S be a set, and f a bounded function on S. Then we define the **sup norm**

$$\|f\|_S = \|f\| = \sup_{z \in S} |f(z)|,$$

where sup means least upper bound. It is a norm in the sense that for two functions f, g we have $\|f + g\| \leq \|f\| + \|g\|$, and for any number c we have $\|cf\| = |c|\,\|f\|$. Also $f = 0$ if and only if $\|f\| = 0$.

Let $\{f_n\}$ ($n = 1, 2, \ldots$) be a sequence of functions on S. We shall say that this sequence **converges uniformly** on S if there exists a function f on S satisfying the following property. Given ϵ, there exists N such that if $n \geq N$, then

$$\|f_n - f\| < \epsilon.$$

We say that $\{f_n\}$ is a **Cauchy sequence** (for the sup norm), if given ϵ, there exists N such that if $m, n \geq N$, then

$$\|f_n - f_m\| < \epsilon.$$

In this case, for each $z \in S$, the sequence of complex numbers

$$\{f_n(z)\}$$

converges, because for each $z \in S$, we have the inequality

$$|f_n(z) - f_m(z)| \leq \|f_n - f_m\|.$$

Theorem 2.1. *If a sequence $\{f_n\}$ of functions on S is Cauchy, then it converges uniformly.*

Proof. For each $z \in S$, let

$$f(z) = \lim_{n \to \infty} f_n(z).$$

Given ϵ, there exists N such that if $m, n \geq N$, then

$$|f_n(z) - f_m(z)| < \epsilon, \qquad \text{for all} \quad z \in S.$$

Let $n \geq N$. Given $z \in S$ select $m \geq N$ sufficiently large (depending on z) such that

$$|f(z) - f_m(z)| < \epsilon.$$

Then

$$
\begin{aligned}
|f(z) - f_n(z)| &\leq |f(z) - f_m(z)| + |f_m(z) - f_n(z)| \\
&< \epsilon + \|f_m - f_n\| \\
&< 2\epsilon.
\end{aligned}
$$

This is true for any z, and therefore $\|f - f_n\| < 2\epsilon$, which proves the theorem.

Remark. It is immediately seen that if the functions f_n in the theorem are bounded, then the limiting function f is also bounded.

Consider a series of functions, $\sum f_n$. Let

$$s_n = \sum_{k=1}^{n} f_k = f_1 + f_2 + \cdots + f_n$$

be the partial sum. We say that the **series converges uniformly** if the sequence of partial sums $\{s_n\}$ converges uniformly.

A series $\sum f_n$ is said to converge **absolutely** if for each $z \in S$ the series

$$\sum |f_n(z)|$$

converges.

The next theorem is sometimes called the **comparison test**.

Theorem 2.2. *Let $\{c_n\}$ be a sequence of real numbers ≥ 0, and assume that*

$$\sum c_n$$

converges. Let $\{f_n\}$ *be a sequence of functions on* S *such that* $\|f_n\| \leq c_n$ *for all* n. *Then* $\sum f_n$ *converges uniformly and absolutely.*

Proof. Say $m \leq n$. We have an estimate for the difference of partial sums,

$$\|s_n - s_m\| \leq \sum_{k=m+1}^{n} \|f_k\| \leq \sum_{k=m+1}^{n} c_k.$$

The assumption that $\sum c_k$ converges implies at once the uniform convergence of the partial sums. The argument also shows that the convergence is absolute.

Theorem 2.3. *Let* S *be a set of complex numbers, and let* $\{f_n\}$ *be a sequence of continuous functions on* S. *If this sequence converges uniformly, then the limit function* f *is also continuous.*

Proof. You should already have seen this theorem some time during a calculus course. We reproduce the proof for convenience. Let $\alpha \in S$. Select n so large that $\|f - f_n\| < \epsilon$. For this choice of n, using the continuity of f_n at α, select δ such that whenever $|z - \alpha| < \delta$ we have

$$|f_n(z) - f_n(\alpha)| < \epsilon.$$

Then

$$|f(z) - f(\alpha)| \leq |f(z) - f_n(z)| + |f_n(z) - f_n(\alpha)| + |f_n(\alpha) - f(\alpha)|.$$

The first and third term on the right are bounded by $\|f - f_n\| < \epsilon$. The middle term is $< \epsilon$. Hence

$$|f(z) - f(\alpha)| < 3\epsilon,$$

and our theorem is proved.

We now consider the power series, where the functions f_n are

$$f_n(z) = a_n z^n,$$

with complex numbers a_n.

Theorem 2.4. *Let* $\{a_n\}$ *be a sequence of complex numbers, and let* r *be a number* > 0 *such that the series*

$$\sum |a_n| r^n$$

converges. Then the series $\sum a_n z^n$ converges absolutely and uniformly for $|z| \leq r$.

Proof. Special case of the comparison test.

Example. For any $r > 0$, the series

$$\sum z^n/n!$$

converges absolutely and uniformly for $|z| \leq r$. Indeed, let

$$c_n = r^n/n!.$$

Then

$$\frac{c_{n+1}}{c_n} = \frac{r^{n+1}}{(n+1)!} \frac{n!}{r^n} = \frac{r}{n+1}.$$

Take $n \geq 2r$. Then the right-hand side is $\leq 1/2$. Hence for all n sufficiently large, we have

$$c_{n+1} \leq \tfrac{1}{2} c_n.$$

Therefore there exists some positive integer n_0 such that

$$c_n \leq C/2^{n-n_0},$$

for some constant C and all $n \geq n_0$. We may therefore compare our series with a geometric series to get the absolute and uniform convergence.

The series

$$\exp(z) = \sum_{n=0}^{\infty} z^n/n!$$

therefore defines a continuous function for all values of z. Similarly, the series

$$\sin z = z - \frac{z^3}{3!} + \frac{z^5}{5!} - \cdots = \sum_{n=0}^{\infty} (-1)^n \frac{z^{2n+1}}{(2n+1)!}$$

and

$$\cos z = 1 - \frac{z^2}{2!} + \frac{z^4}{4!} - \cdots = \sum_{n=0}^{\infty} (-1)^n \frac{z^{2n}}{(2n)!}$$

converge absolutely and uniformly for all $|z| \leq r$. They give extensions of the sine and cosine functions to the complex numbers. We shall see later that $\exp(z) = e^z$ as defined in Chapter I, and that these series define the unique analytic functions which coincide with the usual exponential, since, and cosine functions, respectively, when z is real.

Theorem 2.5. *Let $\sum a_n z^n$ be a power series. If it does not converge absolutely for all z, then there exists a number r such that the series converges absolutely for $|z| < r$ and does not converge absolutely for $|z| > r$.*

Proof. Suppose that the series does not converge absolutely for all z. Let r be the least upper bound of those numbers $s \geq 0$ such that

$$\sum |a_n| s^n$$

converges. Then $\sum |a_n| |z|^n$ diverges if $|z| > r$, and converges if $|z| < r$ by the comparison test, so our assertion is obvious.

The number r in Theorem 2.5 is called the **radius of convergence** of the power series. If the power series converges absolutely for all z, then we say that its **radius of convergence** is **infinity**. When the radius of convergence is 0, then the series converges absolutely only for $z = 0$.

If a power series has a non-zero radius of convergence, then it is called a **convergent** power series. If D is a disc centered at the origin and contained in the disc $D(0, r)$, where r is the radius of convergence, then we say that the power series **converges on** D.

The radius of convergence can be determined in terms of the coefficients. Let t_n be a sequence of real numbers. We recall that a **point of accumulation** of this sequence is a number t such that, given ϵ, there exist infinitely many indices n such that

$$|t_n - t| < \epsilon.$$

In other words, infinitely many points of the sequence lie in a given interval centered at t. An elementary property of real numbers asserts that every bounded sequence has a point of accumulation (Weierstrass–Bolzano theorem).

Assume now that $\{t_n\}$ is a bounded sequence. Let S be the set of points of accumulation, so that S looks like Fig. 1. We define the **limit superior**, lim sup, of the sequence to be

$$\lambda = \limsup t_n = \text{least upper bound of } S.$$

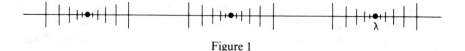

Figure 1

Then the reader will verify at once that λ is itself a point of accumulation of the sequence, and is therefore the largest such point. Furthermore, λ has the following properties:

Given ϵ, there exists only finitely many n such that $t_n \geq \lambda + \epsilon$. There exist infinitely many n such that

$$t_n \geq \lambda - \epsilon.$$

Proof. If there were infinitely many n such that $t_n \geq \lambda + \epsilon$, then the sequence $\{t_n\}$ would have a point of accumulation

$$\geq \lambda + \epsilon > \lambda,$$

contrary to assumption. On the other hand, since λ itself is a point of accumulation, given the ϵ-interval about λ, there has to be infinitely many n such that t_n lies in this ϵ-interval, thus proving the second assertion.

We leave it to the reader to verify that *if a number λ has the above properties, then it is in the* lim sup *of the sequence.*

For convenience, if $\{t_n\}$ is not bounded from above, we define its lim sup to be infinity, written ∞.

As an exercise, you should prove:

Let $\{t_n\}$, $\{s_n\}$ be sequences of real numbers ≥ 0. Let

$$t = \limsup t_n \qquad and \qquad s = \limsup s_n.$$

Then

$$\limsup(t_n + s_n) \leq t + s.$$

If $t \neq 0$, then

$$\limsup(t_n s_n) \leq ts.$$

If $\lim t_n$ exists, then $t = \lim\limits_{n \to \infty} t_n$.
$$\lim\limits_{n \to \infty}$$

This last statement says that if the sequence has an ordinary limit, then that limit is the lim sup of the sequence.

The second statement is often applied in case one sequence has a lim sup, and the other sequence has a limit $\neq 0$. The hypothesis $t \neq 0$ is made only to allow the possibility that $s = \infty$, in which case ts is understood to be ∞. If $s \neq \infty$, and $t \neq \infty$, then it is true unrestrictedly that

$$\lim \sup(t_n s_n) = ts.$$

Theorem 2.6. *Let $\sum a_n z^n$ be a power series, and let r be its radius of convergence. Then*

$$\frac{1}{r} = \lim \sup |a_n|^{1/n}.$$

If $r = 0$, this relation is to be interpreted as meaning that the sequence $\{|a_n|^{1/n}\}$ is not bounded. If $r = \infty$, it is to be interpreted as meaning that $\lim \sup |a_n|^{1/n} = 0$.

Proof. Suppose first that $r \neq 0, \infty$. Let $0 < s < r$. Since the series

$$\sum |a_n| s^n$$

converges, it follows that $|a_n| s^n \to 0$ as $n \to \infty$. Hence there is a number C such that

$$|a_n| s^n \leq C \qquad \text{for all } n.$$

Then

$$|a_n|^{1/n} \leq \frac{C^{1/n}}{s}.$$

But $\lim_{n \to \infty} C^{1/n} = 1$. Hence $\lim \sup |a_n|^{1/n} \leq 1/s$. This is true for every positive number $s < r$, and hence we conclude that

$$\lim \sup |a_n|^{1/n} \leq 1/r.$$

Conversely, let $\lim \sup |a_n|^{1/n} = t$ be a real number, and suppose $t < 1/r$. Let $\epsilon > 0$ be such that $t + \epsilon < 1/r$. For all but a finite number of n we have

$$|a_n|^{1/n} \leq t + \epsilon.$$

Let t_1 be a number such that $t + \epsilon < t_1 < 1/r$. Then

$$|a_n|^{1/n} \frac{1}{t_1} \leq \frac{t + \epsilon}{t_1} < 1 \qquad \text{for all but a finite number of } n.$$

Then omitting a finite number of n in the sum, we get

$$\sum |a_n| \frac{1}{t_1^n} \leq \sum \left(\frac{t + \epsilon}{t_1} \right)^n < \infty$$

so $\sum a_n(1/t_1^n)$ converges. But $1/t_1 > r$, contradicting the fact that r is the radius of convergence. This proves the theorem in the case $r \neq 0, \infty$. We leave the cases $r = 0$ and ∞ to the reader.

Corollary 2.7. *If $\lim |a_n|^{1/n} = t$ exists, then $r = 1/t$.*

Proof. If the limit exists, then t is the only point of accumulation of the sequence $|a_n|^{1/n}$, and the theorem states that $t = 1/r$.

Corollary 2.8. *Suppose that $\sum a_n z^n$ has a radius of convergence greater than 0. Then there exist positive numbers C, A such that*

$$|a_n| \leq CA^n \qquad \text{for all } n.$$

Proof. Let $0 < s < r$, and let $A = 1/s$ at the beginning of the proof of the theorem.

Example. Let us determine the radius of convergence of the series

$$\sum nz^n.$$

We have

$$\lim n^{1/n} = 1.$$

Hence the radius of convergence is $1/1 = 1$.

Example. The radius of convergence of the series

$$\sum n! \, z^n$$

is 0. Indeed, you should be able to show by elementary calculus that

$$n! \geq n^n e^{-n},$$

and therefore

$$(n!)^{1/n} \geq n/e.$$

This is unbounded as $n \to \infty$, so the radius of convergence is 0.

Example. We can see once more that the radius of convergence of the series

$$\sum \frac{1}{n!} z^n$$

is infinity. Indeed,

$$(1/n!)^{1/n} \leq \frac{e}{n},$$

and the right-hand side has the limit 0 as $n \to \infty$. Hence the radius of convergence is $1/0 = \infty$.

Example. **The binomial series.** Let α be any complex number $\neq 0$. Define the **binomial coefficients** as usual,

$$\binom{\alpha}{n} = \frac{\alpha(\alpha - 1) \cdots (\alpha - n + 1)}{n!},$$

and the **binomial series**

$$B_\alpha(T) = \sum \binom{\alpha}{n} T^n.$$

By convention,

$$\binom{\alpha}{0} = 1.$$

The radius of convergence of the binomial series is ≥ 1.

Proof. Let $s = |\alpha|$. We have the estimate

$$\left| \binom{\alpha}{n} \right| \leq \frac{s(1 + s)(2 + s) \cdots (n - 1 + s)}{1 \cdot 2 \cdots n}$$

$$\leq s(1 + s)\left(1 + \frac{s}{2}\right) \cdots \left(1 + \frac{s}{n}\right).$$

It will suffice to prove that the n-th root of the right-hand side approaches 1 as n becomes large. We know that

$$\lim_{n \to \infty} s^{1/n} = 1.$$

We have

$$0 \leq \log\left[(1 + s)\left(1 + \frac{s}{2}\right) \cdots \left(1 + \frac{s}{n}\right)\right]^{1/n}$$

$$= \frac{1}{n}\left[\log(1 + s) + \cdots + \log\left(1 + \frac{s}{n}\right)\right]$$

$$\leq \frac{1}{n}\left(s + \frac{s}{2} + \cdots + \frac{s}{n}\right)$$

$$\leq \frac{s}{n}(1 + \log n).$$

The right-hand side approaches 0 as n becomes large. This proves the log of the desired limit is 0, whence the desired limit is 1. Hence the desired estimate for the radius of convergence has been proved.

Warning. Let r be the radius of convergence of the series $f(z)$. *Nothing* has been said about possible convergence if $|z| = r$. Many cases can occur concerning convergence or non-convergence on this circle, but we shall not worry about such refined questions in this course.

EXERCISES II §2

1. Let $|\alpha| < 1$. Express the sum of the geometric series

$$\sum_{n=1}^{\infty} \alpha^n$$

in its usual simple form.

2. Let r be a real number, $0 \leq r < 1$. Show that the series

$$\sum_{n=0}^{\infty} r^n e^{in\theta}$$

converges (θ is real). Express that series in simple terms using the usual formula for a geometric series. Do the same for

$$\sum_{n=-\infty}^{\infty} r^{|n|} e^{in\theta}.$$

3. Show that the usual power series for $\log(1 + z)$ or $\log(1 - z)$ from elementary calculus converges absolutely for $|z| < 1$.

4. Determine the radius of convergence for the following power series.

(a) $\sum n^n z^n$

(b) $\sum z^n/n^n$

(c) $\sum 2^n z^n$

(d) $\sum (\log n)^2 z^n$

(e) $\sum 2^{-n} z^n$

(f) $\sum n^2 z^n$

(g) $\sum \dfrac{n!}{n^n} z^n$

(h) $\sum \dfrac{(n!)^3}{(3n)!} z^n$

5. Let $f(z) = \sum a_n z^n$ have radius of convergence $r > 0$. Show that the following series have the same radius of convergence:

(a) $\sum n a_n z^n$

(b) $\sum n^2 a_n z^n$

(c) $\sum n^d a_n z^n$ for any positive integer d

(d) $\sum_{n \geq 1} n a_n z^{n-1}$

6. Let a, b be two complex numbers, and assume that b is not equal to any integer ≤ 0. Determine the radius of convergence of the series

$$\sum \frac{a(a + 1)(a + 2) \cdots (a + n)}{b(b + 1)(b + 2) \cdots (b + n)} z^n.$$

7. Let $\sum a_n$ be a convergent series of complex numbers. Prove that the power series $\sum a_n z^n$ is uniformly convergent on the domain of z such that

$$\pi - \lambda < \arg(z - 1) < \pi + \lambda \qquad \text{and} \qquad |z - 1| \leq \delta,$$

where $0 < \lambda < \pi/2$ and $0 < \delta < 2 \cos \lambda$.

8. Let $\{a_n\}$ be a decreasing sequence of positive numbers approaching 0. Prove that the power series $\sum a_n z^n$ is uniformly convergent on the domain of z such that

$$|z| \leq 1 \qquad \text{and} \qquad |z - 1| \geq \delta,$$

where $\delta > 0$.
[*Hint*: For the above two problems, use summation by parts.]

9. Let $\sum a_n z^n$ and $\sum b_n z^n$ be two power series, with radius of convergence r and s, respectively. What can you say about the radius of convergence of the series:

(a) $\sum (a_n + b_n) z^n$

(b) $\sum a_n b_n z^n$?

10. Let $\{a_n\}$ be the sequence of real numbers defined by the conditions:

$$a_0 = 1, a_1 = 2, \qquad \text{and} \qquad a_n = a_{n-1} + a_{n-2} \qquad \text{for} \quad n \geq 2.$$

Determine the radius of convergence of the power series

$$\sum_{n=0}^{\infty} a_n z^n.$$

[*Hint*: What is the general solution of a difference equation? Cf. Exercise 5 of §1.]

II §3. Relations Between Formal and Convergent Series

Sums and Products

Let $f = f(T)$ and $g = g(T)$ be formal power series. We may form their formal product and sum, $f + g$ and fg. If f converges absolutely for some complex number z, then we have the value $f(z)$, and similarly for $g(z)$.

Theorem 3.1. *If f, g are power series which converge absolutely on the disc $D(0, r)$, then $f + g$ and fg also converge absolutely on this disc. If α is a complex number, αf converges absolutely on this disc, and we have*

$$(f + g)(z) = f(z) + g(z), \qquad (fg)(z) = f(z)g(z),$$
$$(\alpha f)(z) = \alpha \cdot f(z)$$

for all z in the disc.

Proof. We give the proof for the product, which is the hardest. Let

$$f = \sum a_n T^n \qquad \text{and} \qquad g = \sum b_n T^n,$$

so that

$$fg = \sum c_n T^n, \qquad \text{where} \quad c_n = \sum_{k=0}^{n} a_k b_{n-k}.$$

Let $0 < s < r$. We know that there exists a positive number C such that for all n,

$$|a_n| \leq C/s^n \qquad \text{and} \qquad |b_n| \leq C/s^n.$$

Then

$$|c_n| \leq \sum_{k=0}^{n} |a_k b_{n-k}| \leq (n + 1) \frac{C}{s^k} \frac{C}{s^{n-k}} = \frac{(n + 1)C^2}{s^n}.$$

Therefore

$$|c_n|^{1/n} \leq \frac{(n + 1)^{1/n} C^{2/n}}{s}.$$

But $\lim_{n \to \infty} (n + 1)^{1/n} C^{1/n} = 1$. Hence

$$\limsup |c_n|^{1/n} \leq 1/s.$$

This is true for every $s < r$. It follows that $\limsup |c_n|^{1/n} \leq 1/r$, thereby proving that the formal product converges absolutely on the same disc. We have also shown that the series of positive terms

$$\sum_{n=0}^{\infty} \sum_{k=0}^{n} |a_k| \, |b_{n-k}| \, |z|^n$$

converges.

Let

$$f_N(T) = a_0 + a_1 T + \cdots + a_N T^N,$$

and similarly, let $g_N(T)$ be the polynomial consisting of the terms of order $\leq N$ in the power series for g. Then

$$f(z) = \lim_N f_N(z) \quad \text{and} \quad g(z) = \lim_N g_N(z).$$

Furthermore,

$$|(fg)_N(z) - f_N(z)g_N(z)| \leq \sum_{n=N+1}^{\infty} \sum_{k=0}^{N} |a_k| \, |b_{n-k}| \, |z|^n.$$

In view of the convergence proved above, for N sufficiently large the right-hand side is arbitrarily small, and hence

$$f(z)g(z) = \lim_N f_N(z)g_N(z) = (fg)(z),$$

thereby proving the theorem for the product.

The previous theorem shows that a formal power series determines a function on the disc of absolute convergence. We can raise the question: If two formal power series f, g give rise to the same function on some neighborhood of 0, are they equal as formal power series? Subtracting g from f, this amounts to asking: If a power series determines the zero function on some disc centered at the origin, is it the zero series, i.e. are all its coefficients equal to 0? The answer is yes. In fact, more is true.

Theorem 3.2. *Let $f(T) = \sum a_n T^n$ be a non-constant power series, having a non-zero radius of convergence. If $f(0) = 0$, then there exists $s > 0$ such that $f(z) \neq 0$ for all z with $|z| \leq s$, and $z \neq 0$.*

Proof. We can write

$$f(z) = a_m z^m + \text{higher terms}, \quad \text{and} \quad a_m \neq 0$$
$$= a_m z^m (1 + b_1 z + b_2 z^2 + \cdots)$$
$$= a_m z^m (1 + h(z)),$$

where $h(z) = b_1 z + b_2 z^2 + \cdots$ is a power series having a non-zero radius of convergence, and zero constant term. For all sufficiently small $|z|$, the value $|h(z)|$ is small, and hence

$$1 + h(z) \neq 0.$$

If $z \neq 0$, then $a_m z^m \neq 0$. This proves the theorem.

As a consequence of Theorem 3.2, we see that there exists at most one power series

$$f(T) = \sum a_n T^n$$

which has a non-zero radius of convergence, and such that for some interval $[-\epsilon, \epsilon]$ we have

$$f(x) = e^x \quad \text{for all } x \text{ in } [-\epsilon, \epsilon].$$

This proves the uniqueness of any power series extension of the exponential function to all complex numbers. Similarly, one has the uniqueness of the power series extending the sine and cosine functions.

Furthermore, let $\exp(z) = \sum z^n/n!$. Then

$$\exp(iz) = \sum (iz)^n/n!.$$

Summing over even n and odd n, we find that

$$\exp(iz) = C(z) + iS(z),$$

where $C(z)$ and $S(z)$ are the power series for the cosine and sine of z, respectively. Hence $e^{i\theta}$ for real θ coincides with $\exp(i\theta)$ as given by the power series expansion. Thus we see that the exponential function e^z defined in Chapter I has the same values as the function defined by the usual power series $\exp(z)$. From now on, we make no distinction between e^z and $\exp(z)$.

Theorem 3.2 also allows us to conclude that any polynomial relation between the elementary functions which have a convergent Taylor expansion at the origin also holds for the extension of these functions as complex power series.

Example. We can now conclude that

$$\sin^2 z + \cos^2 z = 1,$$

where $\sin z = S(z)$, $\cos z = C(z)$ are defined by the usual power series. Indeed,

$$S(z)^2 + C(z)^2$$

is a power series with infinite radius of convergence, taking the value 1 for all real z. Theorem 3.2 implies that there is at most one series having this property, and that is the series 1, as desired. It would be disagreeable to show directly that the formal power series for the sine and cosine satisfy this relation. It is easier to do it through elementary calculus as above.

Example. Let m be a positive integer. We have seen in §2 that the binomial series

$$B(z) = \sum \binom{\alpha}{n} z^n$$

with $\alpha = 1/m$ has a radius of convergence ≥ 1, and thus converges absolutely for $|z| < 1$. By elementary calculus, we have

$$B(x)^m = 1 + x$$

when x is real, and $|x| < 1$ (or even when $|x|$ is sufficiently small). Therefore $B(T)^m$ is the unique formal power series such that

$$B(x)^m = 1 + x$$

for all sufficiently small real x, and therefore we conclude that

$$B(T)^m = 1 + T.$$

In this manner, we see that we can take m-th roots

$$(1 + z)^{1/m}$$

by the binomial series when $|z| < 1$.

Quotients

In our discussion of formal power series, besides the polynomial relations, we dealt with quotients and also composition of series. We still
have to relate these to the convergent case. It will be convenient to
introduce a simple notation to estimate power series.

Let $f(T) = \sum a_n T^n$ be a power series. Let

$$\varphi(T) = \sum c_n T^n$$

be a power series with **real coefficients** $c_n \geq 0$. We say that f is **dominated
by** φ, and write

$$f \prec \varphi \qquad \text{or} \qquad f(T) \prec \varphi(T),$$

if $|a_n| \leq c_n$ for all n. It is clear that if φ, ψ are power series with real co-
efficients ≥ 0 and if

$$f \prec \varphi, \qquad g \prec \psi,$$

then

$$f + g \prec \varphi + \psi \qquad \text{and} \qquad fg \prec \varphi\psi.$$

Theorem 3.3. *Suppose that f has a non-zero radius of convergence, and
non-zero constant term. Let g be the formal power series which is inverse
to f, that is, $fg = 1$. Then g also has a non-zero radius of convergence.*

Proof. Multiplying f by some constant, we may assume without loss
of generality that the constant term is 1, so we write

$$f = 1 + a_1 T + a_2 T^2 + \cdots = 1 - h(T),$$

where $h(T)$ has constant term equal to 0. By Corollary 2.8, we know that
there exists a number $A > 0$ such that

$$|a_n| \leq A^n, \qquad n \geq 1.$$

(We can take $C = 1$ by picking A large enough.) Then

$$\frac{1}{f(T)} = \frac{1}{1 - h(T)} = 1 + h(T) + h(T)^2 + \cdots.$$

But

$$h(T) \prec \sum_{n=1}^{\infty} A^n T^n = \frac{AT}{1 - AT}.$$

Therefore $1/f(T) = g(T)$ satisfies

$$g(T) \prec 1 + \frac{AT}{1 - AT} + \frac{(AT)^2}{(1 - AT)^2} + \cdots = \frac{1}{1 - \dfrac{AT}{1 - AT}}.$$

But

$$\frac{1}{1 - \dfrac{AT}{1 - AT}} = (1 - AT)(1 + 2AT + (2AT)^2 + \cdots)$$

$$\prec (1 + AT)(1 + 2AT + (2AT)^2 + \cdots).$$

Therefore $g(T)$ is dominated by a product of power series having non-zero radius of convergence, whence $g(T)$ itself has a non-zero radius of convergence, as was to be shown.

Composition of Series

Theorem 3.4. *Let*

$$f(z) = \sum_{n \geq 0} a_n z^n \qquad and \qquad h(z) = \sum_{n \geq 1} b_n z^n$$

be convergent power series, and assume that the constant term of h is 0. Assume that $f(z)$ is absolutely convergent for $|z| \leq r$, with $r > 0$, and that $s > 0$ is a number such that

$$\sum |b_n| s^n \leq r.$$

Let $g = f(h)$ be the formal power series obtained by composition,

$$g(T) = \sum_{n \geq 0} a_n \left(\sum_{k=1}^{\infty} b_k T^k \right)^n.$$

Then g converges absolutely for $|z| \leq s$, and for such z,

$$g(z) = f(h(z)).$$

Proof. Let $g(T) = \sum c_n T^n$. Then $g(T)$ is dominated by the series

$$g(T) \prec \sum_{n=0}^{\infty} |a_n| \left(\sum_{k=1}^{\infty} |b_k| T^k \right)^n$$

and by hypothesis, the series on the right converges absolutely for $|z| \leqq s$, so $g(z)$ converges absolutely for $|z| \leqq s$. Let

$$f_N(T) = a_0 + a_1 T + \cdots + a_{N-1} T^{N-1}$$

be the polynomial of degree $\leqq N - 1$ beginning the power series f. Then

$$f(h(T)) - f_N(h(T)) \prec \sum_{n=N}^{\infty} |a_n| \left(\sum_{k=1}^{\infty} |b_k| T^k \right)^n,$$

and $f(h(T)) = g(T)$ by definition. By the absolute convergence we conclude: Given ϵ, there exists N_0 such that if $N \geqq N_0$ and $|z| \leqq s$, then

$$|g(z) - f_N(h(z))| < \epsilon.$$

Since the polynomials f_N converge uniformly to the function f on the closed disc of radius r, we can pick N_0 sufficiently large so that for $N \geqq N_0$ we have

$$|f_N(h(z)) - f(h(z))| < \epsilon.$$

This proves that

$$|g(z) - f(h(z))| < 2\epsilon,$$

for every ϵ, whence $g(z) - f(h(z)) = 0$, thereby proving the theorem.

Example. Let m be a positive integer, and let $h(z)$ be a convergent power series with zero constant term. Then we can form the m-th root

$$(1 + h(z))^{1/m}$$

by the binomial expansion, and this m-th root is a convergent power series whose m-th power is $1 + h(z)$.

EXERCISES II §3

1. Define $\log(1 + z)$ for $|z| < 1$ by the usual power series. Prove that

$$\exp(\log(1 + z)) = 1 + z.$$

2. Show that there exists $r > 0$ such that if $|z| < r$ then

$$\log \exp z = z.$$

3. Prove that for all complex z we have

$$\cos z = \frac{e^{iz} + e^{-iz}}{2} \quad \text{and} \quad \sin z = \frac{e^{iz} - e^{-iz}}{2i}.$$

4. Show that the only complex numbers z such that $\sin z = 0$ are $z = k\pi$, where k is an integer. State and prove a similar statement for $\cos z$.

5. Find the power series expansion of $f(z) = 1/(z + 1)(z + 2)$, and find the radius of convergence.

6. The **Legendre polynomials** can be defined as the coefficients $P_n(\alpha)$ of the series expansion of

$$f(z) = \frac{1}{(1 - 2\alpha z + z^2)^{1/2}}$$

$$= 1 + P_1(\alpha)z + P_2(\alpha)z^2 + \cdots + P_n(\alpha)z^n + \cdots.$$

Calculate the first four Legendre polynomials.

II §4. Analytic Functions

So far we have looked at power series expansions at the origin. Let f be a function defined in some neighborhood of a point z_0. We say that f is **analytic** at z_0 if there exists a power series

$$\sum_{n=0}^{\infty} a_n(z - z_0)^n$$

and some $r > 0$ such that the series converges absolutely for $|z - z_0| < r$, and such that for such z, we have

$$f(z) = \sum_{n=0}^{\infty} a_n(z - z_0)^n.$$

Suppose f is a function on an open set U. We say that f is **analytic on** U if f is analytic at every point of U.

In the light of the uniqueness theorem for power series, Theorem 3.2, we see that the above power series expressing f in some neighborhood of z_0 is uniquely determined. We have

$$f(z_0) = 0 \quad \text{if and only if} \quad a_0 = 0.$$

A point z_0 such that $f(z_0) = 0$ is called a **zero** of f. Instead of saying that f is analytic at z_0, we also say that f **has a power series expansion at** z_0 (meaning that the values of $f(z)$ for z near z_0 are given by an absolutely convergent power series as above).

If S is an arbitrary set, not necessarily open, it is useful to make the convention that a function is **analytic** on S if it is the restriction of a analytic function on an open set containing S. This is useful, for instance, when S is a closed disc.

The theorem concerning sums, products, quotients, and composites of convergent power series now immediately imply:

If f, g are analytic on U, so are $f + g$, fg. Also f/g is analytic on the open subset of $z \in U$ such that $g(z) \neq 0$.

If $g: U \to V$ is analytic and $f: V \to \mathbf{C}$ is analytic, then $f \circ g$ is analytic.

For this last assertion, we note that if $z_0 \in U$ and $g(z_0) = w_0$, so

$$g(z) = w_0 + \sum_{n \geq 1} b_n(z - z_0)^n,$$

then $g(z) - w_0$ is represented by a power series $h(z - z_0)$ without constant term, so that Theorem 3.4 applies: We can "substitute"

$$f(g(z)) = \sum a_n(g(z) - w_0)^n$$

to get the power series representation for $f(g(z))$ in a neighborhood of z_0.

The next theorem, although easy to prove, requires being stated. It gives us in practice a way of finding a power series expansion for a function at a point.

Theorem 4.1. *Let $f(z) = \sum a_n z^n$ be a power series whose radius of convergence is r. Then f is analytic on the open disc $D(0, r)$.*

Proof. We have to show that f has a power series expansion at an arbitrary point z_0 of the disc, so $|z_0| < r$. Let $s > 0$ be such that $|z_0| + s < r$. We shall see that f can be represented by a convergent power series at z_0, converging absolutely on a disc of radius s centered at z_0.

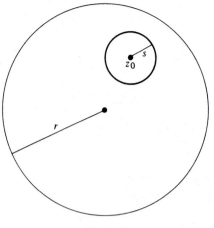

Figure 2

We write

$$z = z_0 + (z - z_0)$$

so that

$$z^n = (z_0 + (z - z_0))^n.$$

Then

$$f(z) = \sum_{n=0}^{\infty} a_n \left(\sum_{k=0}^{n} \binom{n}{k} z_0^{n-k} (z - z_0)^k \right),$$

and the series of positive terms

$$\sum |a_n| (|z_0| + |z - z_0|)^n = \sum_{n=0}^{\infty} |a_n| \left[\sum_{k=0}^{n} \binom{n}{k} |z_0|^{n-k} |z - z_0|^k \right]$$

converges. Then we can interchange the order of summation, to get

$$f(z) = \sum_{k=0}^{\infty} \left[\sum_{n=k}^{\infty} a_n \binom{n}{k} z_0^{n-k} \right] (z - z_0)^k,$$

which converges absolutely also, as was to be shown.

Example. Let us find the terms of order ≤ 3 in the power series expansion of the function

$$f(z) = z^2/(z + 2)$$

at the point $z_0 = 1$. We write

$$z = 1 + (z - 1), \qquad z + 2 = 3 + (z - 1).$$

Let $\equiv$ denote congruence mod z^4 (so disregard terms of order > 3). Then

$$z^2 = 1 + 2(z - 1) + (z - 1)^2$$

$$z + 2 = 3\left(1 + \frac{1}{3}(z - 1)\right)$$

$$\frac{1}{z + 2} = \frac{1}{3} \cdot \frac{1}{1 + \frac{1}{3}(z - 1)}$$

$$= \frac{1}{3}\left(1 - \frac{1}{3}(z - 1) + \frac{1}{3^2}(z - 1)^2 - \frac{1}{3^3}(z - 1)^3 + \cdots\right).$$

Hence

$$\frac{z^2}{z + 2} \equiv (1 + 2(z - 1) + (z - 1)^2)$$

$$\times \frac{1}{3}\left(1 - \frac{1}{3}(z - 1) + \frac{1}{3^2}(z - 1)^2 - \frac{1}{3^3}(z - 1)^3\right)$$

$$= \frac{1}{3}\left[1 + \frac{5}{3}(z - 1) + \left(\frac{1}{3} + \frac{1}{3^2}\right)(z - 1)^2\right.$$

$$\left. + \left(-\frac{1}{3} + \frac{2}{3^2} - \frac{1}{3^3}\right)(z - 1)^3\right].$$

These are the desired terms of the expansion.

Remark.   Making a translation, the theorem shows that if f has a power series expansion on a disc $D(z_0, r)$, that is,

$$f(z) = \sum a_n(z - z_0)^n$$

for $|z - z_0| < r$, then f is analytic on this disc.

II §5. The Inverse and Open Mapping Theorems

Let f be a function on an open set U, and let $f(U) = V$. We shall say that f is a **analytic isomorphism** if V is open and there exists a analytic function

$$g : V \to U$$

such that $f \circ g = \text{id}_V$ and $g \circ f = \text{id}_U$, in other words, f and g are inverse functions to each other.

We say that f is a **local analytic isomorphism** at a point z_0 if there exists an open set U containing z_0 such that f is a analytic isomorphism on U.

Theorem 5.1. *Let f be analytic at z_0, and suppose that in the power series expansion*

$$f(z) = a_0 + a_1(z - z_0) + \cdots$$

we have $a_1 \neq 0$. Then f is a local analytic isomorphism at z_0.

Proof. After making a translation if necessary, we may assume that $z_0 = 0, f(z_0) = 0$, so that

$$f(z) = a_1 z + \cdots, \quad \text{and} \quad a_1 \neq 0.$$

We shall first prove that f is injective in some neighborhood of the origin, in other words, if $z \neq w$, then $f(z) \neq f(w)$ whenever z, w are sufficiently close to the origin. We have

$$\frac{f(z) - f(w)}{z - w} = a_1 + a_2 \frac{z^2 - w^2}{z - w} + a_3 \frac{z^3 - w^3}{z - w} + \cdots,$$

and we can estimate,

$$\left| \frac{z^n - w^n}{z - w} \right| = \left| \sum_{k=0}^{n-1} z^k w^{n-k-1} \right| \leq n r^{n-1}$$

if $|z|, |w| \leq r$. If $f(z) = f(w)$, we obtain

$$|a_1| \leq \sum_{n=2}^{\infty} |a_n| n r^{n-1}.$$

For r sufficiently small, the right-hand side is small (why?), thus giving a contradiction, and proving the injectivity.

We must now prove that f can be inverted locally by a convergent power series. We find first a formal inverse for $f(T)$. We seek a power series

$$g(T) = \sum_{n=1}^{\infty} b_n T^n$$

with $b_1 \neq 0$ such that

$$f(g(T)) = T.$$

The solution to this problem is given by solving the equations in terms of the coefficients of the power series

$$a_1 g(T) + a_2 g(T^2) + \cdots = T,$$

and these equations are of the form

$$a_1 b_n + P_n(a_2, \ldots, a_n, b_1, \ldots, b_{n-1}) = 0,$$

where P_n is a polynomial with positive integer coefficients (generalized binomial coefficients). In fact, one sees at once that

$$P_n(a_2, \ldots, a_n, b_1, \ldots, b_{n-1})$$
$$= a_2 P_{2,n}(b_1, \ldots, b_{n-1}) + \cdots + a_n P_{n,n}(b_1, \ldots, b_{n-1}),$$

where again $P_{k,n}$ is a polynomial with positive coefficients, but we won't need this more explicit description of the polynomials P_n. In this manner we can solve recursively for the coefficients

$$b_1, b_2, \ldots$$

since b_n appears linearly with coefficient $a_1 \neq 0$ in these equations, and the other terms do not contain b_n. This shows that a formal inverse exists.

We must now show that $g(z)$ is absolutely convergent on some disc. To simplify the number of symbols used, we assume that $a_1 = 1$. This loses no generality, because if we find a convergent inverse power series for $a_1^{-1}f(z)$, we immediately get the convergent inverse power series for $f(z)$ itself.

Let

$$f^*(T) = T - \sum_{n \geq 2} a_n^* T^n$$

be a power series with a_n^* real ≥ 0 such that $|a_n| \leq a_n^*$ for all n. Let $\varphi(T)$ be the formal inverse of $f^*(T)$, so

$$\varphi(T) = \sum_{n \geq 1} c_n T^n, \qquad c_1 = 1.$$

Then we have

$$c_n - P_n(a_1^*, \ldots, a_n^*, c_1, \ldots, c_{n-1}) = 0$$

with those same polynomials P_n as before. By induction, it therefore follows that c_n is real ≥ 0, and also that

$$|b_n| \leq c_n.$$

It suffices therefore to pick the series f^* so that it has an easily computed formal inverse φ which is easily verified to have a positive radius of convergence.

It is now a simple matter to carry out this idea, and we pick for f^* a geometric series. There exists $A > 0$ such that for all n we have

$$|a_n| \leq A^n.$$

(We can omit a constant C in front of A^n by picking A sufficiently large.) Then

$$f^*(T) = T - \sum_{n \geq 2} A^n T^n = T - \frac{A^2 T^2}{1 - AT}.$$

The power series $\varphi(T)$ is such that $f^*(\varphi(T)) = T$, namely

$$\varphi(T) - \frac{A^2 \varphi(T)^2}{1 - A\varphi(T)} = T,$$

which is equivalent with the quadratic equation

$$(A^2 + A)\varphi(T)^2 - (1 + AT)\varphi(T) + T = 0.$$

This equation has the solution

$$\varphi(T) = \frac{1 + AT - \sqrt{(1 + AT)^2 - 4T(A^2 + A)}}{2(A^2 + A)}.$$

The expression under the radical sign is of the form

$$(1 + AT)^2 \left(1 - \frac{4T(A^2 + A)}{(1 + AT)^2}\right)$$

and its square root is given by

$$(1 + AT)\left(1 - \frac{4T(A^2 + A)}{(1 + AT)^2}\right)^{1/2}.$$

We use the binomial expansion to find the square root of a series of the form $1 + h(T)$ when $h(T)$ has zero constant term. It is now clear that

$\varphi(T)$ is obtained by composition of convergent power series, and hence has a non-zero radius of convergence. This proves that the power series $g(T)$ also converges.

Let U be an open disc centered at the origin, contained in the disc of convergence of f, and such that f is injective on U. We can find such U by the first part of the proof. Let V be an open disc centered at the origin contained in the disc of convergence of g, and such that $g(V) \subset U$. We can find V by the continuity of g. Then

$$f(U) \supset f(g(V)) = V.$$

Let U_0 be the set of $z \in U$ such that $f(z) \in V$. Then U_0 is open by the continuity of f, and

$$f: U_0 \to V$$

is both injective and surjective, so bijective. Since both f and its inverse map g are analytic, we see that f is a local analytic isomorphism, thereby proving the theorem.

In §7 we shall prove that if f has a power series expansion

$$f(z) = \sum a_n z^n$$

then its derivative is given by the usual formula

$$f'(z) = \sum n a_n z^{n-1}.$$

Therefore

$$a_1 = f'(0),$$

and instead of assuming $a_1 \neq 0$ in Theorem 5.1, we could have assumed equivalently that $f'(0) \neq 0$.

There are (at least) four ways of proving the inverse function theorem.

1. The way we have just gone through, by estimating the formal inverse to show that it converges.

2. Reproducing the real variable proof for real functions of class C^1. By the contraction principle, (shrinking lemma), one first shows that the map is locally surjective, and one constructs a local inverse, which is

shown to be differentiable, and whose derivative satisfies, for $w = f(z)$, the relation

$$g'(w) = 1/f'(z).$$

The reader should be able to copy the proof from any standard book on analysis, (certainly from my *Undergraduate Analysis*). In §7 we shall prove that $f'(z)$ is also holomorphic, and is given by the "obvious" power series. It then follows that g is also analytic.

3. Assuming the theorem for C^∞ real functions. One can show (and we shall do so later when we discuss the real aspects of an analytic function) that an analytic function is C^∞, as a function of (x, y), writing

$$z = x + iy.$$

The hypothesis $f'(z_0) \neq 0$ (namely $a_1 \neq 0$) is then seen to amount to the property that the Jacobian of the real function of two variables has non-zero determinant, whence f has a C^∞ inverse locally by the real theorem. It is then an easy matter to show by the chain rule that this inverse satisfies the Cauchy–Riemann equations, and is therefore holomorphic, whence analytic by the theory which follows Cauchy's theorem.

4. Giving an argument based on more complex function theory, and carried out in Theorem 1.5 of Chapter VI.

All four methods are important, and are used in various contexts in analysis, both of functions of one variable, and functions of several variables.

Let U be an open set and let f be a function on U. We say that f is an **open mapping** if for every open subset U' of U the image $f(U')$ is open.

Theorem 5.1 shows that the particular type of function considered there, i.e. with non-zero first coefficient in the power series expansion, is locally open. We shall now consider arbitrary analytic functions, first at the origin.

Let

$$f(z) = \sum a_n z^n$$

be a convergent non-constant power series, and let $m = \operatorname{ord} f$, so that

$$f(z) = a_m z^m + \text{higher terms}, \qquad a_m \neq 0.$$
$$= a_m z^m (1 + h(z)),$$

where $h(z)$ is convergent, and has zero constant term. Let a be a complex number such that $a^m = a_m$. Then we can write $f(z)$ in the form

$$f(z) = (az(1 + h_1(z)))^m,$$

where $h_1(z)$ is a convergent power series with zero constant term, obtained from the binomial expansion

$$(1 + h(z))^{1/m} = 1 + h_1(z),$$

and

$$f_1(z) = az(1 + h_1(z)) = az + azh_1(z)$$

is a power series whose coefficient of z is $a \neq 0$. Theorem 5.1 therefore applies to $f_1(z)$, which is therefore locally open at the origin. We have

$$f(z) = f_1(z)^m.$$

Let U be an open disc centered at the origin on which f_1 converges. Then $f_1(U)$ contains an open disc V. The image of V under the map

$$w \mapsto w^m$$

is a disc. Hence $f(U)$ contains an open disc centered at the origin.

Theorem 5.2. *Let f be analytic on an open set U, and assume that for each point of U, f is not constant on a given neighborhood of that point. Then f is an open mapping.*

Proof. We apply the preceding discussion to the power series expansion of f at a point of U, so the proof is obvious in the light of what we have already done.

The construction in fact yielded the following statement which it is worthwhile extracting as a theorem.

Theorem 5.3. *Let f be analytic at a point z_0,*

$$f(z) = a_0 + \sum_{n=m}^{\infty} a_n(z - z_0)^n,$$

with $m \geqq 1$ *and* $a_m \neq 0$. *Then there exists a local analytic isomorphism* φ *at* 0 *such that*

$$f(z) = a_0 + \varphi(z - z_0)^m.$$

We interpret Theorem 5.3 as follows. Let

$$\psi : U \to V$$

be a analytic isomorphism. We write $w = \psi(z)$. We may view ψ as a change of coordinates, from the coordinate z to the coordinate w. In Theorem 5.3 we may therefore write

$$f(z) = a_0 + w^m,$$

where $w = \varphi(z - z_0)$. The expression for f in terms of the coordinate w is therefore much simpler than in terms of the coordinate z.

We also get a criterion for a function to have an analytic inverse on a whole open set.

Theorem 5.4. *Let* f *be analytic on an open set* U, *and assume that* f *is injective. Let* $V = f(U)$ *be its image. Then* $f : U \to V$ *is an analytic isomorphism.*

Proof. The function f between U and V is bijective, so we can define an inverse mapping $g : V \to U$. Let z_0 be a point of U, and let the power series expansion of f at z_0 be as in Theorem 5.3. If $m > 1$ then we see that f cannot be injective, because the m-th power function in a neighborhood of the origin is not injective (it wraps the disc m times around). Hence $m = 1$, and Theorem 5.1 now shows that the inverse function g is analytic at $g(z_0)$. This proves the theorem. Again if we use the results of §7, we conclude:

If f *is analytic and injective on* U, *then* $f'(z) \neq 0$ *for all* z *in* U.

II §6. The Local Maximum Modulus Principle

This principle is an immediate application of the open mapping theorem, and so we give it here, to emphasize its direct dependence with the preceding section. On the other hand, we wait for a later chapter for less basic applications mostly for psychological reasons. We want to alternate the formal operations with power series and the techniques which will arise from Cauchy's theorem. The later chapter could logically be read almost in its entirety after the present section, however.

We say that a function f is **locally constant** at a point z_0 if there exists an open set D (or a disc) containing z_0 such that f is constant on D.

Theorem 6.1. *Let f be analytic on an open set U. Let $z_0 \in U$ be a maximum for $|f|$, that is,*

$$|f(z_0)| \geqq |f(z)|, \qquad \text{for all} \quad z \in U.$$

Then f is locally constant at z_0.

Proof. The function f has a power series expansion at z_0,

$$f(z) = a_0 + a_1(z - z_0) + \cdots.$$

If f is not the constant $a_0 = f(z_0)$, then by Theorem 5.2 we know that f is an open mapping in a neighborhood of z_0, and therefore the image of f contains a disc $D(a_0, s)$ of radius $s > 0$, centered at a_0. Hence the set of numbers $|f(z)|$, for z in a neighborhood of z_0, contains an open interval around a_0. Hence

$$|f(z_0)| = |a_0|$$

cannot be a maximum for f, a contradiction which proves the theorem.

Corollary. 6.2. *Let f be analytic on an open set U, and let $z_0 \in U$ be a maximum for the real part $\text{Re} \, f$, that is,*

$$\text{Re} \, f(z_0) \geq \text{Re} \, f(z), \qquad \text{for all} \quad z \in U.$$

Then f is locally constant at z_0.

Proof. The function $e^{f(z)}$ is analytic on U, and if

$$f(z) = u(z) + iv(z)$$

is the expression of f in terms of its real and imaginary parts, then

$$|e^{f(z)}| = e^{u(z)}.$$

Hence a maximum for $\text{Re} \, f$ is also a maximum for $|e^{f(z)}|$, and the corollary follows from the theorem.

The theorem is often applied when f is analytic on an open set U and is continuous at the boundary of U. Then a maximum for $|f(z)|$ necessarily occurs on the boundary of U. For this one needs that U is connected, and the relevant form of the theorem will be proved as Theorem 1.3 of the next chapter.

We shall give here one more example of the power of the maximum modulus principle, and postpone to a later chapter some of the other applications.

Theorem 6.3. *Let*

$$f(z) = a_0 + a_1 z + \cdots + a_d z^d$$

be a polynomial, not constant, and say $a_d \neq 0$. Then f has some complex zero, i.e. number z_0 such that $f(z_0) = 0$.

Proof. Suppose otherwise, so that $1/f(z)$ is defined for all z, and defines an analytic function. Writing

$$f(z) = a_d z^d \left(\frac{a_0}{a_d z^d} + \frac{a_1 z}{a_d z^d} + \cdots + 1 \right),$$

one sees that

$$\lim_{|z| \to \infty} 1/f(z) = 0.$$

Let α be some complex number such that $f(\alpha) \neq 0$. Pick a positive number R large enough such that $|\alpha| < R$, and if $|z| \geq R$, then

$$\frac{1}{|f(z)|} < \frac{1}{|f(\alpha)|}.$$

Let S be the closed disc of radius R centered at the origin. Then S is closed and bounded, and $1/|f(z)|$ is continuous on S, whence has a maximum on S, say at z_0. By construction, this point z_0 cannot be on the boundary of the disc, and must be an interior point. By the maximum modulus principle, we conclude that $1/f(z)$ is locally constant at z_0. This is obviously impossible since f itself is not locally constant, say from the expansion

$$f(z) = b_0 + b_1(z - z_0) + \cdots + b_d(z - z_0)^d,$$

with suitable coefficients $b_0, \ldots, b_d$ and $b_d \neq 0$. This proves the theorem.

II §7. Differentiation of Power Series

Let $D(0, r)$ be a disc of radius $r > 0$. A function f on this disc for which there exists a power series $\sum a_n z^n$ having a radius of convergence $\geq r$ and such that

$$f(z) = \sum a_n z^n$$

for all z in the disc is said to **admit a power series expansion on this disc**. We shall now see that such a function is holomorphic, and that its derivative is given by the "obvious" power series.

Indeed, define the **formal derived series** to be

$$\sum n a_n z^{n-1} = a_1 + 2a_2 z + 3a_3 z^2 + \cdots.$$

Theorem 7.1. *If $f(z) = \sum a_n z^n$ has radius of convergence r, then:*

(i)   *The series $\sum n a_n z^{n-1}$ has the same radius of convergence.*
(ii) *The function f is holomorphic on $D(0, r)$, and its derivative is equal to $\sum n a_n z^{n-1}$.*

Proof. By Theorem 2.6, we have

$$\limsup |a_n|^{1/n} = 1/r.$$

But

$$\limsup |n a_n|^{1/n} = \limsup n^{1/n} |a_n|^{1/n}.$$

Since $\lim n^{1/n} = 1$, the sequences

$$|n a_n|^{1/n} \qquad \text{and} \qquad |a_n|^{1/n}$$

have the same lim sup, and therefore the series $\sum a_n z^n$ and $\sum n a_n z^n$ have the same radius of convergence. Then

$$\sum n a_n z^{n-1} \qquad \text{and} \qquad \sum n a_n z^n$$

converge absolutely for the same values of z, so the first part of the theorem is proved.

As to the second, let $|z| < r$, and $\delta > 0$ be such that $|z| + \delta < r$. We consider complex numbers h such that

$$|h| < \delta.$$

We have:

$$f(z + h) = \sum a_n(z + h)^n$$
$$= \sum a_n(z^n + nz^{n-1}h + h^2 P_n(z, h)),$$

where $P_n(z, h)$ is a polynomial in z and h, with positive integer coefficients, in fact

$$P_n(z, h) = \sum_{k=2}^{n} \binom{n}{k} h^{k-2} z^{n-k}.$$

Note that we have the estimate:

$$|P_n(z, h)| \leq \sum_{k=2}^{n} \binom{n}{k} \delta^{k-2} |z|^{n-k} = P_n(|z|, \delta).$$

Subtracting series, we find

$$f(z + h) - f(z) - \sum na_n z^{n-1} h = h^2 \sum a_n P_n(z, h),$$

and since the series on the left is absolutely convergent, so is the series on the right. We divide by h to get

$$\frac{f(z + h) - f(z)}{h} - \sum na_n z^{n-1} = h \sum a_n P_n(z, h).$$

For $|h| < \delta$, we have the estimate

$$|\sum a_n P_n(z, h)| \leq \sum |a_n| |P_n(z, h)|$$
$$\leq \sum |a_n| P_n(|z|, \delta).$$

This last expression is fixed, independent of h. Hence

$$|h \sum a_n P_n(z, h)| \leq |h| \sum |a_n| P_n(|z|, \delta).$$

As h approaches 0, the right-hand side approaches 0, and therefore

$$\lim_{h \to 0} |h \sum a_n P_n(z, h)| = 0.$$

This proves that f is differentiable, and that its derivative at z is given by the series $\sum na_n z^{n-1}$, as was to be shown.

Remark. Conversely, we shall see after Cauchy's theorem that a function which is differentiable admits power series expansions—a very remarkable fact, characteristic of complex differentiability.

From the theorem, we see that the k-th derivative of f is given by the series

$$f^{(k)}(z) = k!\, a_k + h_k(z),$$

where h_k is a power series without constant term. Therefore we obtain the standard expression for the coefficients of the power series in terms of the derivatives, namely

$$a_n = \frac{f^{(n)}(0)}{n!}.$$

If we deal with the expansion at a point z_0, namely

$$f(z) = a_0 + a_1(z - z_0) + a_2(z - z_0)^2 + \cdots,$$

then we find

$$a_n = \frac{f^{(n)}(z_0)}{n!}.$$

It is utterly trivial that the formally integrated series

$$\sum_{n=0}^{\infty} \frac{a_n}{n+1} z^{n+1}$$

has radius of convergence at least r, because its coefficients are smaller in absolute value than the coefficients of f. Since the derivative of this integrated series is exactly the series for f, it follows from Theorem 6.1 that the integrated series has the same radius of convergence as f.

Let f be a function on an open set U. If g is a holomorphic function on U such that $g' = f$, then g is called a **primitive** for f. We see that a function which has a power series expansion on a disc always has a primitive on that disc. In other words, an analytic function has a local primitive at every point.

Example. The function $1/z$ is analytic on the open set U consisting of the plane from which the origin has been deleted. Indeed, for $z_0 \neq 0$, we have the power series expansion

$$\frac{1}{z} = \frac{1}{z_0 + z - z_0} = \frac{1}{z_0} \frac{1}{(1 + (z - z_0)/z_0)}$$

$$= \frac{1}{z_0} \left(1 - \frac{1}{z_0}(z - z_0) + \cdots \right)$$

converging on some disc $|z - z_0| < r$. Hence $1/z$ has a primitive on such a disc, and this primitive may be called $\log z$.

EXERCISES II §7

In Exercises 1 through 5, also determine the radius of convergence of the given series.

1. Let

$$f(z) = \sum \frac{z^{2n}}{(2n)!}$$

Prove that $f''(z) = f(z)$.

2. Let

$$f(z) = \sum_{n=0}^{\infty} \frac{z^{2n}}{(n!)^2}.$$

Prove that

$$z^2 f''(z) + z f'(z) = 4z^2 f(z).$$

3. Let

$$f(z) = z - \frac{z^3}{3} + \frac{z^5}{5} - \frac{z^7}{7} + \cdots.$$

Show that $f'(z) = 1/(z^2 + 1)$.

4. Let

$$J(z) = \sum_{n=0}^{\infty} \frac{(-1)^n}{(n!)^2} \left(\frac{z}{2} \right)^{2n}.$$

Prove that

$$z^2 J''(z) + z J'(z) + z^2 J(z) = 0.$$

5. For any positive integer k, let

$$J_k(z) = \sum_{n=0}^{\infty} \frac{(-1)^n}{n!\,(n+k)!} \left(\frac{z}{2}\right)^{2n+k}.$$

Prove that

$$z^2 J_k''(z) + z J_k'(z) + (z^2 - k^2) J_k(z) = 0.$$

6. **Linear Differential Equations.** Prove:

Theorem. *Let $a_0(z), \ldots, a_k(z)$ be analytic functions in a neighborhood of 0. Assume that $a_0(0) \neq 0$. Given numbers $c_0, \ldots, c_{k-1}$, there exists a unique analytic function f at 0 such that*

$$D^n f(0) = c_n \quad for \quad n = 0, \ldots, k-1$$

and such that

$$a_0(z) D^k f(z) + a_1(z) D^{k-1} f(z) + \cdots + a_k(z) f(z) = 0.$$

[*Hint*: First you may assume $a_0(z) = 1$ (why?). Then solve for f by a formal power series. Then prove this formal series converges.]

7. **Ordinary Differential Equations.** Prove:

Theorem. *Let g be analytic at 0. There exists a unique analytic function f at 0 satisfying*

$$f(0) = 0, \quad and \quad f'(z) = g(f(z)).$$

[*Hint*: Again find a formal solution, and then prove that it converges.]

CHAPTER III

Cauchy's Theorem, First Part

III §1. Holomorphic Functions on Connected Sets

Let $[a, b]$ be a closed interval of real numbers. By a **curve** γ (defined on this interval) we mean a function

$$\gamma \colon [a, b] \to \mathbf{C}$$

which we assume to be of class C^1.

Figure 1

We recall what this means. We write

$$\gamma(t) = \gamma_1(t) + i\gamma_2(t),$$

where γ_1 is the real part of γ, and γ_2 is its imaginary part. For instance the curve

$$\gamma(\theta) = \cos \theta + i \sin \theta, \qquad 0 \leq \theta \leq 2\pi,$$

is the unit circle. **Of class** C^1 means that the functions $\gamma_1(t)$, $\gamma_2(t)$ have continuous derivatives in the ordinary sense of calculus. We have drawn a curve in Fig. 1. Thus a curve is a parametrized curve. We call $\gamma(a)$ the

beginning point, and $\gamma(b)$ the **end point** of the curve. By a **point on the curve** we mean a point w such that $w = \gamma(t)$ for some t in the interval of definition of γ.

We define the derivative $\gamma'(t)$ in the obvious way, namely

$$\gamma'(t) = \gamma'_1(t) + i\gamma'_2(t).$$

It is easily verified as usual that the rules for the derivative of a sum, product, quotient, and chain rule are valid in this case, and we leave this as an exercise. In fact, prove systematically the following statements:

Let $F: [a, b] \to \mathbf{C}$ and $G: [a, b] \to \mathbf{C}$ be complex valued differentiable functions, defined on the same interval. Then:

$$(F + G)' = F' + G',$$

$$(FG)' = FG' + F'G,$$

$$(F/G)' = (GF' - FG')/G^2$$

(this quotient rule being valid only on the set where $G(t) \neq 0$).

Let $\psi: [c, d] \to [a, b]$ be a differentiable function. Then $\gamma \circ \psi$ is differentiable, and

$$(\gamma \circ \psi)'(t) = \gamma'(\psi(t))\psi'(t),$$

as illustrated on Fig. 2(i).

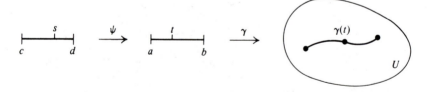

Figure 2(i)

Finally suppose γ is a curve in an open set U and

$$f: U \to \mathbf{C}$$

is a holomorphic function. Then the composite $f \circ \gamma$ is differentiable (as a function of the real variable t) and

$$(f \circ \gamma)'(t) = f'(\gamma(t))\gamma'(t),$$

as illustrated on Fig. 2(ii).

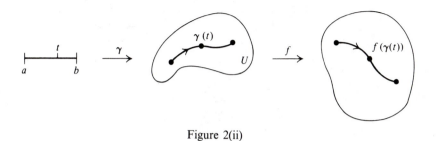

Figure 2(ii)

It is technically convenient to deal with a generalization of curves. By a **path** we shall mean a sequence of curves,

$$\gamma = \{\gamma_1, \gamma_2, \ldots, \gamma_n\}$$

(so each curve γ_j is C^1) such that the end point of γ_j is equal to the beginning point of γ_{j+1}. If γ_j is defined on the interval $[a_j, b_j]$, this means that

$$\gamma_j(b_j) = \gamma_{j+1}(a_j).$$

We have drawn a path on Fig. 3, where z_j is the end point of γ_j. We call $\gamma_1(a_1)$ the **beginning point** of γ, and $\gamma_n(b_n)$ the **end point** of γ. The path is said to **lie in an open set** U if each curve γ_j lies in U, i.e. for each t, the point $\gamma_j(t)$ lies in U.

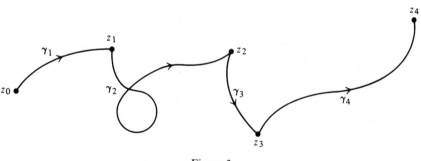

Figure 3

We define an open set U to be **connected** if given two points α and β in U, there exists a path $\{\gamma_1, \ldots, \gamma_n\}$ in U such that α is the beginning point of γ_1 and β is the end point of γ_n; in other words, if there exists a path in U which joins α to β. In Fig. 4 we have drawn an open set which is not connected. In Fig. 5 we have drawn a connected open set. (The definition of connected applies of course equally well to a set which is not

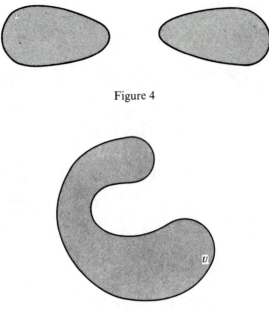

Figure 4

Figure 5

necessarily open. It is usually called **pathwise connected**, but for open sets, this coincides with another possible definition. See the appendix of §2.)

Theorem 1.1. *Let U be a connected open set, and let f be a holomorphic function on U. If $f' = 0$ then f is constant.*

Proof. Let α, β be two points in U, and suppose first that γ is a curve joining α to β, so that

$$\gamma(a) = \alpha \qquad \text{and} \qquad \gamma(b) = \beta.$$

The function

$$t \mapsto f(\gamma(t))$$

is differentiable, and by the chain rule, its derivative is

$$f'((t))\gamma'(t) = 0.$$

Hence this function is constant, and therefore

$$f(\alpha) = f((a)) = f((b)) = f(\beta).$$

Next, suppose that $\gamma = \{\gamma_1, \ldots, \gamma_n\}$ is a path joining α to β, and let z_j be the end point of γ_j, putting

$$z_0 = \alpha, \qquad z_n = \beta.$$

By what we have just proved, we have

$$f(\alpha) = f(z_0) = f(z_1) = f(z_2) = \cdots = f(z_n) = f(\beta),$$

thereby proving the theorem.

If f is a function on an open set U and g is a holomorphic function on U such that $g' = f$, then we call g a **primitive** of f on U. Theorem 1.1 says that on a connected open set, a primitive of f is uniquely determined up to a constant, i.e. if g_1 and g_2 are two primitives, then $g_1 - g_2$ is constant, because the derivative of $g_1 - g_2$ is equal to 0.

In what follows we shall attempt to get primitives by integration. On the other hand, primitives can also be written down directly.

Example. For each integer $n \neq -1$, the function $f(z) = z^n$ has the usual primitive

$$\frac{z^{n+1}}{n+1}.$$

Let S be a set of points, and let $z_0 \in S$. We say that z_0 is **isolated** in S if there exists a disc $D(z_0, r)$ of some radius $r > 0$ such that $D(z_0, r)$ does not contain any point of S other than z_0. We say that S is **discrete** if every point of S is isolated.

Theorem 1.2. *Let U be a connected open set.*

(i) *If f is analytic on U and not constant, then the set of zeros of f on U is discrete.*

(ii) *Let f, g be analytic on U. Let S be a set of points in U which is not discrete (so some point of S is not isolated). Assume that $f(z) = g(z)$ for all z in S. Then $f = g$ on U.*

Proof. We observe that (ii) follows from (i). It suffices to consider the difference $f - g$, which is locally constant, equal to 0 in a neighborhood of z_0. Therefore we set about to prove (i). We know from Theorem 3.2 of the preceding chapter that either f is locally constant and equal to 0 in the neighborhood of a zero z_0, or z_0 is an isolated zero.

Suppose that f is equal to 0 in the neighborhood of some point z_0. We have to prove that $f(z) = 0$ for all $z \in U$. Let S be the set of points

z such that f is equal to 0 in a neighborhood of z. Then S is open. By Theorem 1.6 of the appendix, it will suffice to prove that S is closed in U. Let z_1 be a point in the closure of S in U. Since f is continuous, it follows that $f(z_1) = 0$. If z_1 is not in S, then there exist points of S arbitrarily close to z_1, and by Theorem 3.2 of the preceding chapter, it follows that f is locally equal to 0 in a neighborhood of z_1. Hence in fact $z_1 \in S$, so S is closed in U. This concludes the proof.

Remarks. The argument using open and closed subsets of U applies in very general situations, and shows how to get a global statement on a connected set U knowing only a local property as in Theorem 3.2 of the preceding chapter.

It will be proved in Chapter V, §1 that a function is holomorphic if and only if it is analytic. Thus Theorem 1.2 will also apply to holomorphic functions.

The second part of Theorem 1.2 will be used later in the study of analytic continuation, but we make some comments here in anticipation. Let f be an analytic function defined on an open set U and let g be an analytic function defined on an open set V. Suppose that U and V have a non-empty intersection, as illustrated on Fig. 6. If U, V are connected, and if $f(z) = g(z)$ for all $z \in U \cap V$, i.e. if f and g are equal on the intersection $U \cap V$, then Theorem 1.2 tells us that g is the only possible analytic function on V having this property. In the applications, we shall be interested in extending the domain of definition of an analytic function f, and Theorem 1.2 guarantees the uniqueness of the extended function. We say that g is the **analytic continuation** of f to V.

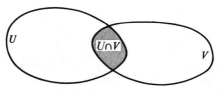

Figure 6

It is also appropriate here to formulate the global version of the **maximum modulus principle**.

Theorem 1.3. *Let U be a connected open set, and let f be an analytic function on U. If $z_0 \in U$ is a maximum point for $|f|$, that is*

$$|f(z_0)| \geqq |f(z)|$$

for all $z \in U$, then f is constant on U.

Proof. By Theorem 6.1 of the preceding chapter, we know that f is locally constant at z_0. Therefore f is constant on U by Theorem 1.2(ii) (compare the constant function and f). This concludes the proof.

Corollary 1.4. *Let U be a connected open set and U^c its closure. Let f be a continuous function on U^c, analytic and non-constant on U. If z_0 is a maximum for f on U^c, that is, $|f(z_0)| \geq |f(z)|$ for all $z \in U^c$, then z_0 lies on the boundary of U^c.*

Proof. This comes from a direct application of Theorem 1.3.

Remark. If U^c is closed and bounded, then a continuous function has a maximum on U^c, so a maximum for f always exists in Corollary 1.4.

Appendix: Connectedness

The purpose of this appendix is to put together a couple of statements describing connectedness in various terms. We start with intervals.

Lemma 1.5. *Let $\gamma: [a, c] \to \mathbf{C}$ be a continuous function. If γ is not constant, then no point in the image of γ is isolated.*

Proof. Suppose the image of γ contains an isolated point $z_1 = \gamma(t_1)$. We let $D(z_1, r)$ be an open disc of radius $r > 0$ centered at z_1 such that $D(z_1, r)$ does not contain any other point of the image of γ. Let:

$$S = \{t, \gamma(t) \in D(z_1, r)\} = \{t, \gamma(t) = z_1\},$$

$$S' = \{t, \gamma(t) \notin D(z_1, r)\}.$$

Then both S, S' are closed and not empty. One of S, S' does not contain c. Say S' does not contain c. Let λ be the least upper bound of S'. Then $\lambda \in S'$ because S' is closed. On the other hand, there exist elements of S arbitrarily close to λ, because $\lambda \neq c$. This contradicts the hypothesis that S is closed, and proves the lemma.

Let A be a set of complex numbers and let T be a subset of A. We say that T is **closed in** A if given $z \in A$ and $z \in T^c$ (the closure of T), then $z \in T$.

Theorem 1.6. *Let U be an open connected set. Let S be a non-empty open subset which is closed in U. Then $S = U$.*

Proof. Suppose $S \neq U$. Let T be the complement of S in U, so T is not empty. Then T is also open and closed in U (verification left to the reader). Let $z_1 \in S$ and $z_2 \in T$. Since U is connected, there is a path

$$\gamma: [a, b] \to U$$

such that $\gamma(a) = z_1$ and $\gamma(b) = z_2$. Let

$$f: U \to \mathbf{R}$$

be the function such that $f(S) = 0$ and $f(T) = 1$. Then f is continuous, and

$$f \circ \gamma: [a, b] \to \{0, 1\}$$

is a continuous function. This contradicts Lemma 1.4, and concludes the proof.

Remark. The converse of Theorem 1.6 is also true, but will not be used. Prove it as an exercise. In this light, Theorem 1.6 gives an equivalent condition for an open set in $\mathbf{C}$ to be connected.

EXERCISE III §1

1. Let U be a bounded open connected set, $\{f_n\}$ a sequence of continuous functions on the closure of U, analytic on U. Assume that $\{f_n\}$ converges uniformly on the boundary of U. Prove that $\{f_n\}$ converges uniformly on U.

III §2. Integrals Over Paths

Let $F: [a, b] \to \mathbf{C}$ be a continuous function.

Write F in terms of its real and imaginary parts, say

$$F(t) = u(t) + iv(t).$$

Define the **indefinite integral** by

$$\int F(t)\, dt = \int u(t)\, dt + i \int v(t)\, dt.$$

Verify that integration by parts is valid (assuming that F' and G' exist and are continuous), namely

$$\int F(t)G'(t)\, dt = F(t)G(t) - \int G(t)F'(t)\, dt.$$

(The proof is the same as in ordinary calculus, from the derivative of a product.)

We define the **integral of F** over $[a, b]$ to be

$$\int_a^b F(t)\, dt = \int_a^b u(t)\, dt + i \int_a^b v(t)\, dt.$$

Thus the integral is defined in terms of the ordinary integrals of the real functions u and v. Consequently, by the fundamental theorem of calculus *the function*

$$t \mapsto \int_a^t F(s)\, ds$$

is differentiable, and its derivative is $F(t)$, because this assertion is true if we replace F by u and v, respectively.

Using simple properties of the integral of real-valued functions, one has the inequality

$$\left| \int_a^b F(t)\, dt \right| \leq \int_a^b |F(t)|\, dt.$$

Work it out as Exercise 11.

Let f be a continuous function on an open set U, and suppose that γ is a curve in U, meaning that all values $\gamma(t)$ lie in U for $a \leq t \leq b$. We define the **integral of f along γ** to be

$$\int_\gamma f = \int_a^b f(\gamma(t))\gamma'(t)\, dt.$$

This is also frequently written

$$\int_\gamma f(z)\, dz.$$

Example 1. Let $f(z) = 1/z$. Let $\gamma(\theta) = e^{i\theta}$. Then

$$\gamma'(\theta) = ie^{i\theta}.$$

We want to find the value of the integral of f over the circle,

$$\int_\gamma \frac{1}{z}\, dz$$

so $0 \leq \theta \leq 2\pi$. By definition, this integral is equal to

$$\int_0^{2\pi} \frac{1}{e^{i\theta}} ie^{i\theta}\, d\theta = i \int_0^{2\pi} d\theta = 2\pi i.$$

As in calculus, we have defined the integral over parametrized curves. In practice, we sometimes describe a curve without giving an explicit parametrization. The context should always make it clear what is meant. Furthermore, one can also easily see that the integral is independent of the parametrization, in the following manner

Let

$$g: [a, b] \to [c, d]$$

be a C^1 function, such that $g(a) = c$, $g(b) = d$, and let

$$\psi: [c, d] \to \mathbf{C}$$

be a curve. Then we may form the composed curve

$$\gamma(t) = \psi(g(t)).$$

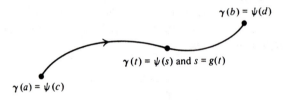

$\gamma(b) = \psi(d)$

$\gamma(t) = \psi(s)$ and $s = g(t)$

$\gamma(a) = \psi(c)$

Figure 7

We find:

$$\int_\gamma f = \int_a^b f(\gamma(t))\gamma'(t)\, dt$$

$$= \int_a^b f(\psi(g(t)))\psi'(g(t))g'(t)\, dt$$

$$= \int_c^d f(\psi(s))\psi'(s)\, ds$$

$$= \int_\psi f.$$

Thus the integral of f along the curve is independent of the parametrization.

If $\gamma = \{\gamma_1, \ldots, \gamma_n\}$ is a path, then we define

$$\int_\gamma f = \sum_{i=1}^n \int_{\gamma_i} f,$$

to be the sum of the integrals of f over each curve γ_i of the path.

Theorem 2.1. *Let f be continuous on an open set U, and suppose that f has a primitive g, that is, g is holomorphic and $g' = f$. Let α, β be two points of U, and let γ be a path in U joining α to β. Then*

$$\int_\gamma f = g(\beta) - g(\alpha),$$

and in particular, this integral depends only on the beginning and end point of the path. It is independent of the path itself.

Proof. Assume first that the path is a curve. Then

$$\int_\gamma f(z)\, dz = \int_a^b g'(\gamma(t))\gamma'(t)\, dt.$$

By the chain rule, the expression under the integral sign is the derivative

$$\frac{d}{dt} g(\gamma(t)).$$

Hence by ordinary calculus, the integral is equal to

$$g(\gamma(t))\Big|_a^b = g(\gamma(b)) - g(\gamma(a)),$$

which proves the theorem in this case. In general, if the path consists of curves $\gamma_1,\ldots,\gamma_n$, and z_j is the end point of γ_j, then by the case we have just settled, we find

$$\int_\gamma f = g(z_1) - g(z_0) + g(z_2) - g(z_1) + \cdots + g(z_n) - g(z_{n-1})$$

$$= g(z_n) - g(z_0),$$

which proves the theorem.

Example 2. Let $f(z) = z^3$. Then f has a primitive, $g(z) = z^4/4$. Hence the integral of f from $2 + 3i$ to $1 - i$ over *any path* is equal to

$$\frac{(1-i)^4}{4} - \frac{(2+3i)^4}{4}.$$

Example 3. Let $f(z) = e^z$. Find the integral of f from 1 to $i\pi$ taken over a line segment. Here again $f'(z) = f(z)$, so f has a primitive. Thus the integral is independent of the path and equal to $e^{i\pi} - e^1 = -1 - e$.

By a **closed path**, we mean a path whose beginning point is equal to its end point. We may now give an important example of the theorem:

If f is a continuous function on U admitting a holomorphic primitive g, and γ is any closed path in U, then

$$\int_\gamma f = 0.$$

Example 4. Let $f(z) = z^n$, where n is an integer $\neq -1$. Then for any closed path γ (or any closed path not passing through the origin if n is negative), we have

$$\int_\gamma z^n \, dz = 0.$$

This is true because z^n has the primitive $z^{n+1}/(n+1)$. [When n is negative, we have to assume that the closed path does not pass through the origin, because the function is then not defined at the origin.]

We shall see later that holomorphic functions are analytic. In that case, in the domain of convergence a power series

$$\sum a_n(z - z_0)^n$$

can be integrated term by term, and thus integrals of holomorphic functions are reduced to integrals of polynomials. This is the reason why there is no need here to give further examples.

Theorem 2.2. *Let U be a connected open set, and let f be a continuous function on U. If the integral of f along any closed path in U is equal to 0, then f has a primitive g on U, that is, a function g which is holomorphic such that $g' = f$.*

Proof. Pick a point z_0 in U and define

$$g(z) = \int_{z_0}^z f,$$

where the integral is taken along any path from z_0 to z in U. If γ, η are two such paths, and η^- is the reverse path of η, then $\{\gamma, \eta^-\}$ is a closed path, and hence (cf. Exercise 9) we know that

$$\int_\gamma f = \int_\eta f.$$

Therefore the integral defining g is independent of the path from z_0 to z, and defines the function. We have

$$\frac{g(z + h) - g(z)}{h} = \frac{1}{h} \int_z^{z+h} f(\zeta) \, d\xi,$$

and the integral from z to $z + h$ can be taken along a segment in U from z to $z + h$. Write

$$f(\zeta) = f(z) + \varphi(\zeta),$$

where $\lim_{\zeta \to z} \varphi(\zeta) = 0$ (this can be done by the continuity of f at z). Then

$$\frac{1}{h} \int_z^{z+h} f(\zeta) \, d\zeta = \frac{1}{h} \int_z^{z+h} f(z) \, d\zeta + \frac{1}{h} \int_z^{z+h} \varphi(\zeta) \, d\zeta$$

$$= f(z) + \frac{1}{h} \int_z^{z+h} \varphi(\zeta) \, d \, .$$

The length of the interval from z to $z + h$ is $|h|$. Hence the integral on the right is estimated by (see below, Theorem 2.3)

$$\frac{1}{|h|} |h| \max |\varphi(\zeta)|,$$

where the max is taken for ζ on the interval. This max tends to 0 as $h \to 0$, and this proves the theorem.

Remarks. The reader should recognize Theorems 2.1 and 2.2 as being the exact analogues for (complex) differentiable functions of the standard theorems of advanced calculus concerning the relation between the existence of a primitive (potential function for a vector field), and the independence of the integral (of a vector field) from the path. We shall see later that a holomorphic function is infinitely complex differentiable, and therefore that f itself is analytic.

Let γ be a curve, $\gamma \colon [a, b] \to \mathbf{C}$, assumed of class C^1 as always. The **speed** is defined as usual to be $|\gamma'(t)|$, and the **length** $L(\gamma)$ is defined to be the integral of the speed,

$$L(\gamma) = \int_a^b |\gamma'(t)| \, dt.$$

If $\gamma = \{\gamma_1, \ldots, \gamma_n\}$ is a path, then by definition

$$L(\gamma) = \sum_{i=1}^{n} L(\gamma_i).$$

Let f be a bounded function on a set S. We let $\|f\|$ be the **sup norm**, written $\|f\|_S$ if the reference to S needs to be made for clarity, so that

$$\|f\| = \sup_{z \in S} |f(z)|$$

is the least upper bound of the values $|f(z)|$ for $z \in S$.

Let f be continuous on an open set U. By standard results of elementary real analysis, Theorem 4.3 of Chapter I, the image of a curve or a path γ is closed and bounded, i.e. compact. If the curve is in U, then the function

$$t \mapsto f(\gamma(t))$$

is continuous, and hence f is bounded on the image of γ. By the compactness of the image of γ, we can always find an open subset of U containing γ, on which f is bounded. If γ is defined on $[a, b]$, we let

$$\|f\|_\gamma = \max_{t \in [a,b]} |f(\gamma(t))|.$$

Theorem 2.3. *Let f be a continuous function on U. Let γ be a path in U. Then*

$$\left| \int_\gamma f \right| \leq \|f\|_\gamma L(\gamma).$$

Proof. If γ is a curve, then

$$\left| \int_\gamma f \right| = \left| \int_a^b f(\gamma(t)) \gamma'(t) \, dt \right|$$

$$\leq \int_a^b |f(\gamma(t))| \, |\gamma'(t)| \, dt$$

$$\leq \|f\|_\gamma L(\gamma),$$

as was to be shown. The statement for a path follows by taking an appropriate sum.

Theorem 2.4. *Let $\{f_n\}$ be a sequence of continuous functions on U, converging uniformly to a function f. Then*

$$\lim \int_\gamma f_n = \int_\gamma f.$$

If $\sum f_n$ is a series of continuous functions converging uniformly on U, then

$$\int_\gamma \sum f_n = \sum \int_\gamma f_n.$$

Proof. The first assertion is immediate from the inequality.

$$\left| \int_\gamma f_n - \int_\gamma f \right| \leqq \int_\gamma |f_n - f| \leqq \|f_n - f\| L(\gamma).$$

The second follows from the first because uniform convergence of a series is defined in terms of the uniform convergence of its partial sums,

$$s_n = f_1 + \cdots + f_n.$$

This proves the theorem.

EXERCISES III §2

1. Given an arbitrary point z_0, let C be a circle of radius $r > 0$ centered at z_0, oriented counterclockwise. Find the integral

$$\int_C (z - z_0)^n \, dz$$

for all integers n, positive or negative.

2. Find the integral of $f(z) = e^z$ from -3 to 3 taken along a semicircle. Is this integral different from the integral taken over the line segment between the two points?

3. Sketch the following curves with $0 \leqq t \leqq 1$.
   (a) $\gamma(t) = 1 + it$
   (b) $\gamma(t) = e^{-\pi it}$
   (c) $\gamma(t) = e^{\pi it}$
   (d) $\gamma(t) = 1 + it + t^2$

4. Find the integral of each one of the following functions over each one of the curves in Exercise 3.
   (a) $f(z) = z^3$
   (b) $f(z) = \bar{z}$
   (c) $f(z) = 1/z$

5. Find the integral

$$\int_\gamma ze^{z^2} \, dz$$

(a) from the point i to the point $-i + 2$, taken along a straight line segment, and
(b) from 0 to $1 + i$ along the parabola $y = x^2$.

6. Find the integral

$$\int_\gamma \sin z \, dz$$

from the origin to the point $1 + i$, taken along the parabola

$$y = x^2.$$

7. Let σ be a vertical segment, say parametrized by

$$\sigma(t) = z_0 + itc, \qquad -1 \le t \le 1,$$

where z_0 is a fixed complex number, and c is a fixed real number >0. (Draw the picture.) Let $\alpha = z_0 + x$ and $\alpha' = z_0 - x$, where x is real positive. Find

$$\lim_{x \to 0} \int_\sigma \left(\frac{1}{z - \alpha} - \frac{1}{z - \alpha'} \right) dz.$$

(Draw the picture.)

8. Let $x > 0$. Find the limit:

$$\lim_{B \to \infty} \int_{-B}^{B} \left(\frac{1}{t + ix} - \frac{1}{t - ix} \right) dt.$$

9. Let $\gamma: [a, b] \to \mathbf{C}$ be a curve. Define the **reverse** or **opposite** curve to be

$$\gamma^-: [a, b] \to \mathbf{C}$$

such that $\gamma^-(t) = \gamma(a + b - t)$. Show that

$$\int_{\gamma^-} F = -\int_\gamma F.$$

10. Let $[a, b]$ and $[c, d]$ be two intervals (not reduced to a point). Show that there is a function $g(t) = rt + s$ such that g is strictly increasing, $g(a) = c$ and $g(b) = d$. Thus a curve can be parametrized by any given interval.

11. Let F be a continuous complex-valued function on the interval $[a, b]$. Prove that

$$\left| \int_a^b F(t)\, dt \right| \leq \int_a^b |F(t)|\, dt.$$

[*Hint*: Let $P = [a = a_0, a_1, \ldots, a_n = b]$ be a partition of $[a, b]$. From the definition of integrals with Riemann sums, the integral

$$\int_a^b F(t)\, dt \quad \text{is approximated by the Riemann sum} \quad \sum_{k=0}^{n-1} F(a_k)(a_{k+1} - a_k)$$

whenever $\max(a_{k+1} - a_k)$ is small, and

$$\int_a^b |F(t)|\, dt \quad \text{is approximated by} \quad \sum_{k=0}^{n-1} |F(a_k)|(a_{k+1} - a_k).$$

The proof is concluded by using the triangle inequality.]

III §3. Local Primitive for a Holomorphic Function

Let U be a connected open set, and let f be holomorphic on U. Let $z_0 \in U$. We want to define a primitive for f on some open disc centered at z_0, i.e. locally at z_0. The natural way is to define such a primitive by an integral,

$$g(z) = \int_{z_0}^{z} f(\zeta)\, d\ ,$$

taken along some path from z_0 to z. However, the integral may depend on the path.

It turns out that we may define g locally by using only a special type of path. Indeed, suppose U is a disc centered at z_0. Let $z \in U$. We select for a path from z_0 to z the edges of a rectangle as shown on Fig. 8.

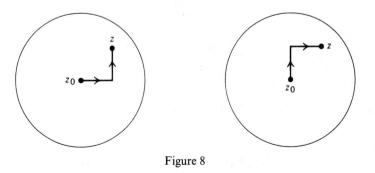

Figure 8

We then have restricted our choice of path to two possible choices as shown. We shall see that we get the same value for the integrals in the two cases. It will be shown afterwards that the integral then gives us a primitive.

By a **rectangle** R we shall mean a rectangle whose sides are vertical or horizontal, and R is meant as the set of points inside and on the boundary of the rectangle, so R is assumed to be closed. The path describing the boundary of the rectangle taken counterclockwise will be also called the **boundary of the rectangle**, and will be denoted by

$$\partial R.$$

If S is an arbitrary set of points, we say that a function f is **holomorphic on** S if it is holomorphic on some open set containing S.

Theorem 3.1 (Goursat). *Let R be a rectangle, and let f be a function holomorphic on R. Then*

$$\int_{\partial R} f = 0.$$

Proof. Decompose the rectangle into four rectangles by bisecting the sides, as shown on Fig. 9.

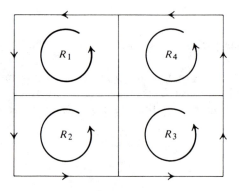

Figure 9

Then

$$\int_{\partial R} f = \sum_{i=1}^{4} \int_{\partial R_i} f.$$

Consequently,

$$\left| \int_{\partial R} f \right| \leq \sum_{i=1}^{4} \left| \int_{\partial R_i} f \right|,$$

and there is one rectangle, say $R^{(1)}$, among R_1, R_2, R_3, and R_4 such that

$$\left| \int_{\partial R^{(1)}} f \right| \geq \frac{1}{4} \left| \int_{\partial R} f \right|.$$

Next we decompose $R^{(1)}$ into four rectangles, again bisecting the sides of $R^{(1)}$ as shown on Fig. 10.

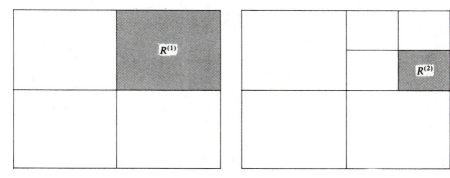

Figure 10

For one of the four rectangles thus obtained, say $R^{(2)}$, we have the similar inequality

$$\left| \int_{\partial R^{(2)}} f \right| \geq \frac{1}{4} \left| \int_{\partial R^{(1)}} f \right|.$$

We continue in this way, to obtain a sequence of rectangles

$$R^{(1)} \supset R^{(2)} \supset R^{(3)} \supset \cdots$$

such that

$$\left| \int_{\partial R^{(n+1)}} f \right| \geq \frac{1}{4} \left| \int_{\partial R^{(n)}} f \right|.$$

Then

$$\left| \int_{\partial R^{(n)}} f \right| \geq \frac{1}{4^n} \left| \int_{\partial R} f \right|.$$

On the other hand, let L_n be the length of $\partial R^{(n)}$. Then

$$L_{n+1} = \frac{1}{2} L_n$$

so that by induction,

$$L_n = \frac{1}{2^n} L_0,$$

where $L_0 =$ length of ∂R.

We contend that the intersection

$$\bigcap_{n=1}^{\infty} R^{(n)}$$

consists of a single point z_0. Since the diameter of $R^{(n)}$ tends to 0 as n becomes large, it is immediate that there is at most one point in the intersection. Let α_n be the center of $R^{(n)}$. Then the sequence $\{\alpha_n\}$ is a Cauchy sequence, because given ϵ, let N be such that the diameter of $R^{(N)}$ is less than ϵ. If $n, m \geq N$, then α_n, α_m lie in $R^{(N)}$ and so

$$|\alpha_n - \alpha_m| \leqq \text{diam } R^{(N)} < \epsilon.$$

Let $z_0 = \lim \alpha_n$. Then z_0 lies in each rectangle, because each rectangle is closed. Hence z_0 lies in the intersection of the rectangles $R^{(N)}$ for $N = 1, 2, \ldots$, as desired.

Since f is differentiable at z_0, there is a disc V centered at z_0 such that for all $z \in V$ we have

$$f(z) = f(z_0) + f'(z_0)(z - z_0) + (z - z_0)h(z),$$

where

$$\lim_{z \to z_0} h(z) = 0.$$

If n is sufficiently large, then $R^{(n)}$ is contained in V, and then

$$\int_{\partial R^{(n)}} f(z)\, dz = \int_{\partial R^{(n)}} f(z_0)\, dz + f'(z_0) \int_{\partial R^{(n)}} (z - z_0)\, dz$$

$$+ \int_{\partial R^{(n)}} (z - z_0)h(z)\, dz.$$

By Examples 1 and 2 of §2, we know that the first two integrals on the right of this equality sign are 0. Hence

$$\int_{\partial R^{(n)}} f = \int_{\partial R^{(n)}} (z - z_0)h(z) \, dz,$$

and we obtain the inequalities

$$\frac{1}{4^n}\left|\int_{\partial R} f\right| \leq \left|\int_{\partial R^{(n)}} f\right| \leq \left|\int_{\partial R^{(n)}} (z - z_0)h(z) \, dz\right|$$

$$\leq \frac{1}{2^n} L_0 \operatorname{diam} R^{(n)} \sup |h(z)|,$$

where the sup is taken for all $z \in R^{(n)}$. But $\operatorname{diam} R^{(n)} = (1/2^n)\operatorname{diam} R$. This yields

$$\left|\int_{\partial R} f\right| \leq L_0 \operatorname{diam} R \sup |h(z)|.$$

The right-hand side tends to 0 as n becomes large, and consequently

$$\int_{\partial R} f = 0,$$

as was to be shown.

We carry out the program outlined at the beginning of the section to find a primitive locally.

Theorem 3.2. *Let U be a disc centered at a point z_0. Let f be continuous on U, and assume that for each rectangle R contained in U we have*

$$\int_{\partial R} f = 0.$$

For each point z_1 in the disc, define

$$g(z_1) = \int_{z_0}^{z_1} f,$$

where the integral is taken along the sides of a rectangle R whose opposite vertices are z_0 and z_1. Then g is analytic on U and is a primitive for f, namely

$$g'(z) = f(z).$$

Proof. We have

$$g(z_1 + h) - g(z_1) = \int_{z_1}^{z_1 + h} f(z)\, dz$$

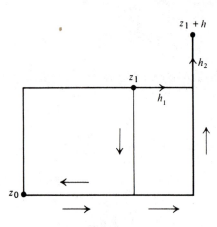

Figure 11

The integral between z_1 and $z_1 + h$ is taken over the bottom side h_1 and vertical side h_2 of the rectangle shown in Fig. 11. Since f is continuous at z_1, there exists a function $\psi(z)$ such that

$$\lim_{z \to z_1} \psi(z) = 0$$

and

$$f(z) = f(z_1) + \psi(z).$$

Then

$$g(z_1 + h) - g(z_1) = \int_{z_1}^{z_1 + h} f(z_1)\, dz + \int_{z_1}^{z_1 + h} \psi(z)\, dz$$

$$= hf(z_1) + \int_{z_1}^{z_1 + h} \psi(z)\, dz.$$

We divide by h and take the limit as $h \to 0$. The length of the path from z_1 to $z_1 + h$ is bounded by $|h_1| + |h_2|$. Hence we get a bound

$$\left| \frac{1}{h} \int_{z_1}^{z_1 + h} \psi(z)\, dz \right| \leq \frac{1}{|h|} (|h_1| + |h_2|) \sup |\psi(z)|,$$

where the sup is taken for z on the path of integration. The expression on the right therefore tends to 0 as $z \to z_1$. Hence

$$\lim_{h \to 0} \frac{g(z_1 + h) - g(z_1)}{h} = f(z_1),$$

as was to be shown.

Knowing that a primitive for f exists on a disc U centered at z_0, we can now conclude that the integral of f along any path between z_0 and z in U is independent of the path, according to Theorem 2.1, and we find:

Theorem 3.3. *Let U be a disc and suppose that f is holomorphic on U. Then f has a primitive on U, and the integral of f along any closed path in U is 0.*

III §4. Another Description of the Integral Along a Path

Knowing the existence of a local primitive for a holomorphic function allows us to describe its integral along a path in a way which makes no use of the differentiability of the path, and would apply to a continuous path as well. We start with curves.

Lemma 4.1. *Let $\gamma: [a, b] \to U$ be a continuous curve in an open set U. Then there is some positive number $r > 0$ such that every point on the curve lies at distance $\leq r$ from the complement of U.*

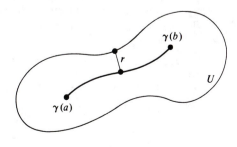

Figure 12

Proof. The image of γ is compact. Consider the function

$$\varphi(t) = \min_{w} |\gamma(t) - w|,$$

where the minimum is taken for all w in the complement of U. This minimum exists because it suffices to consider w lying inside some big circle. Then $\varphi(t)$ is easily verified to be a continuous function of t, whence φ has a minimum on $[a, b]$, and this minimum cannot be 0 because U is open. This proves our assertion.

Let $P = [a_0, \ldots, a_n]$ be a partition of the interval $[a, b]$. We also write P in the form

$$a = a_0 \leq a_1 \leq a_2 \leq \cdots \leq a_n = b.$$

Let $\{D_0, \ldots, D_n\}$ be a sequence of discs. We shall say that this sequence of discs is **connected by the curve along the partition** if D_i contains the image $\gamma([a_i, a_{i+1}])$. The following figure illustrates this.

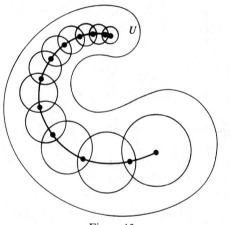

Figure 13

One can always find a partition and such a connected sequence of discs. Indeed, let $\epsilon > 0$ be a positive number such that $\epsilon < r/2$ where r is as in Lemma 4.1. Since γ is uniformly continuous, there exists δ such that if $t, s \in [a, b]$ and $|t - s| < \delta$, then $|\gamma(t) - \gamma(s)| < \epsilon$. We select an integer n and a partition P such that each interval $[a_i, a_{i+1}]$ has length $< \delta$. Then the image $\gamma([a_i, a_{i+1}])$ lies in a disc D_i centered at $\gamma(a_i)$ of radius ϵ, and this disc is contained in U.

Let f be holomorphic on U. Let $\gamma_i : [a_i, b_i] \to U$ be the restriction of γ to the smaller interval $[a_i, b_i]$. Then

$$\int_\gamma f = \sum_{i=0}^{n-1} \int_{\gamma_i} f.$$

Let $\gamma(a_i) = z_i$, and let g_i be a primitive of f on the disc D_i. If each γ_i is of class C^1 then we find:

$$\int_\gamma f = \sum_{i=0}^{n-1} [g_i(z_{i+1}) - g_i(z_i)].$$

Thus even though f may not have a primitive g on the whole open set U, its integral can nevertheless be expressed in terms of local primitives by decomposing the curve as a sum of sufficiently smaller curves. The same formula then applies to a path.

This procedure allows us to define the **integral of f along any continuous curve**; we do not need to assume any differentiability property of the curve. We need only apply the above procedure, but then we must show that the expression

$$\sum_{i=0}^{n-1} [g_i(z_{i+1}) - g_i(z_i)]$$

is independent of the choice of partition of the interval $[a, b]$ and of the choices of the discs D_i containing $\gamma([a_i, a_{i+1}])$. Then this sum can be taken as the definition of the integral

$$\int_\gamma f.$$

The reader interested only in applications may omit the following considerations. First we state formally this independence, repeating the construction.

Lemma 4.2. *Let* $\gamma: [a, b] \to U$ *be a continuous curve. Let*

$$a_0 = a \leq a_1 \leq a_2 \leq \cdots \leq a_n = b$$

be a partition of $[a, b]$ *such that the image* $\gamma([a_i, a_{i+1}])$ *is contained in a disc* D_i, *and* D_i *is contained in* U. *Let* f *be holomorphic on* U *and let* g_i *be a primitive of* f *on* D_i.

Let $z_i = \gamma(a_i)$. *Then the sum*

$$\sum_{i=0}^{n-1} [g_i(z_{i+1}) - g_i(z_i)]$$

is independent of the choices of partitions, discs D_i, *and primitives* g_i *on* D_i *subject to the stated conditions.*

Proof. First let us work with the given partition, but let B_i be another disc containing the image $\gamma([a_i, a_{i+1}])$, and B_i contained in U. Let h_i be a primitive of f on B_i. Then both g_i, h_i are primitives of f on the intersection $B_i \cap D_i$, which is open and connected. Hence there exists a constant C_i such that $g_i = h_i + C_i$ on $B_i \cap D_i$. Therefore the differences are equal:

$$g_i(z_{i+1}) - g_i(z_i) = h_i(z_{i+1}) - h_i(z_i).$$

Thus we have proved that given the partition, the value of the sum is independent of the choices of primitives and choices of discs.

Given two partitions, we can always find a common refinement, as in elementary calculus. Recall that a partition

$$Q = [b_0, \ldots, b_m]$$

is called a **refinement** of the partition P if every point of Q is among the points of P, that is if each b_j is equal to some a_i. Two partitions always have a common refinement, which we obtain by inserting all the points of one partition into the other. Furthermore, we can obtain a refinement of a partition by inserting one point at a time. Thus it suffices to prove that if the partition Q is a refinement of the partition P obtained by inserting one point, then Lemma 4.2 is valid in this case. So we can suppose that Q is obtained by inserting some point c in some interval $[a_k, a_{k+1}]$ for some k, that is Q is the partition

$$[a_0, \ldots, a_k, c, a_{k+1}, \ldots, a_n].$$

We have already shown that given a partition, the value of the sum as in the statement of the lemma is independent of the choice of discs and primitives as described in the lemma. Hence for this new partition Q, we can take the same discs D_i for all the old intervals $[a_i, a_{i+1}]$ when $i \neq k$, and we take the disc D_k for the intervals $[a_k, c]$ and $[c, a_{k+1}]$. Similarly, we take the primitive g_i on D_i as before, and g_k on D_k. Then the sum with respect to the new partition is the same as for the old one, except that the single term

$$g_k(z_{k+1}) - g_k(z_k)$$

is now replaced by two terms

$$g_k(z_{k+1}) - g_k(\gamma(c)) + g_k(\gamma(c)) - g_k(z_k).$$

This does not change the value, and concludes the proof of Lemma 4.2.

For any continuous path $\gamma \colon [a, b] \to U$ we may thus **define**

$$\int_\gamma f = \sum_{i=0}^{n-1} \left[g_i(\gamma(a_{i+1})) - g(\gamma(a_i)) \right]$$

for any partition $[a_0, a_1, \ldots, a_n]$ of $[a, b]$ such that $\gamma([a_i, a_{i+1}])$ is contained in a disc D_i, $D_i \subset U$, and g_i is a primitive of f on D_i. We have just proved that the expression on the right-hand side is independent of the choices made, and we had seen previously that if γ is piecewise C^1 then the expression on the right-hand side gives the same value as the definition used in §2. It is often convenient to have the additional flexibility provided by arbitrary continuous paths.

Remark. The technique of propagating discs along a curve will again be used in the chapter on holomorphic continuation along a curve.

As an application, we shall now see that if two paths lie "close together," and have the same beginning point and the same end point, then the integrals of f along the two paths have the same value. We must define precisely what we mean by "close together". After a reparametrization, we may assume that the two paths are defined over the same interval $[a, b]$. We say that they are **close together** if there exists a partition

$$a = a_0 \leqq a_1 \leqq a_2 \leqq \cdots \leqq a_n = b,$$

and for each $i = 0, \ldots, n - 1$ there exists a disc D_i contained in U such that the images of each segment $[a_i, a_{i+1}]$ under the two paths γ, η are contained in D_i, that is,

$$\gamma([a_i, a_{i+1}]) \subset D_i \qquad \text{and} \qquad \eta([a_i, a_{i+1}]) \subset D_i.$$

Lemma 4.3. *Let γ, η be two continuous paths in an open set U, and assume that they have the same beginning point and the same end point. Assume also that they are close together. Let f be holomorphic on U. Then*

$$\int_\gamma f = \int_\eta f.$$

Proof. We suppose that the paths are defined on the same interval $[a, b]$, and we choose a partition and discs D_i as above. Let g_i be a primitive of f on D_i. Let

$$z_i = \gamma(a_i) \qquad \text{and} \qquad w_i = \eta(a_i).$$

We illustrate the paths and their partition in Fig. 14.

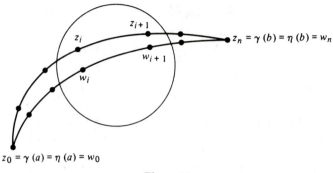

Figure 14

But g_{i+1} and g_i are primitives of f on the connected open set $D_{i+1} \cap D_i$, so $g_{i+1} - g_i$ is constant on $D_{i+1} \cap D_i$. But $D_{i+1} \cap D_i$ contains z_{i+1} and w_{i+1}. Consequently

$$g_{i+1}(z_{i+1}) - g_{i+1}(w_{i+1}) = g_i(z_{i+1}) - g_i(w_{i+1}).$$

Then we find

$$\int_\gamma f - \int_\eta f = \sum_{i=0}^{n-1} [g_i(z_{i+1}) - g_i(z_i) - (g_i(w_{i+1}) - g_i(w_i))]$$
$$= \sum_{i=0}^{n-1} [(g_i(z_{i+1}) - g_i(w_{i+1})) - (g_i(z_i) - g_i(w_i))]$$
$$= g_n(z_n) - g_n(w_n) - (g_0(z_0) - g_0(w_0))$$
$$= 0,$$

because the two paths have the same beginning point $z_0 = w_0$, and the same end point $z_n = w_n$. This proves the lemma.

One can also formulate an analogous lemma for closed paths.

Lemma 4.4. *Let γ, η be closed continuous paths in the open set U, say defined on the same interval $[a, b]$. Assume that they are close together. Let f be holomorphic on U. Then*

$$\int_\gamma f = \int_\eta f.$$

Proof. The proof is the same as above, except that the reason why we find 0 in the last step is now slightly different. Since the paths are closed, we have

$$z_0 = z_n \quad \text{and} \quad w_0 = w_n,$$

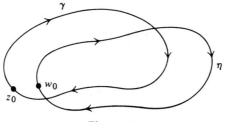

Figure 15

as illustrated in Fig. 15. The two primitives g_{n-1} and g_0 differ by a constant on some disc contained in U and containing z_0, w_0. Hence the last expression obtained in the proof of Lemma 4.3 is again equal to 0, as was to be shown.

III §5. The Homotopy Form of Cauchy's Theorem

Let γ, η be two paths in an open set U. After a reparametrization if necessary, we assume that they are defined over the same interval $[a, b]$. We shall say that γ is **homotopic** to η if there exists a continuous function

$$\psi: [a, b] \times [c, d] \to U$$

defined on a rectangle $[a, b] \times [c, d]$, such that

$$\psi(t, c) = \gamma(t) \qquad \text{and} \qquad \psi(t, d) = \eta(t)$$

for all $t \in [a, b]$.

For each number s in the interval $[c, d]$, we may view the function ψ_s such that

$$\psi_s(t) = \psi(t, s)$$

as a continuous curve, defined on $[a, b]$, and we may view the family of continuous curves ψ_s as a deformation of the path γ to the path η. The picture is drawn on Fig. 16. The paths have been drawn with the same end points because that's what we are going to use in practice. Formally, we say that the homotopy ψ **leaves the end points fixed** if we have

$$\psi(a, s) = \gamma(a) \qquad \text{and} \qquad \psi(b, s) = \gamma(b)$$

for all values of s in $[c, d]$. *In the sequel it will be always understood that when we speak of a homotopy of paths having the same end points, then the homotopy leaves the end points fixed.*

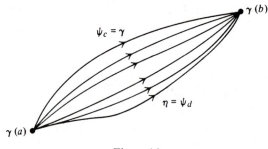

Figure 16

Similarly, when we speak of a homotopy of closed paths, **we assume always that each path ψ_s is a closed path**. These additional requirements are now regarded as part of the definition of homotopy and will not be repeated each time.

Theorem 5.1. *Let γ, η be paths in an open set U having the same beginning point and the same end point. Assume that they are homotopic in U. Let f be holomorphic on U. Then*

$$\int_\gamma f = \int_\eta f.$$

Theorem 5.2. *Let γ, η be closed paths in U, and assume that they are homotopic in U. Let f be holomorphic on U. Then*

$$\int_\gamma f = \int_\eta f.$$

In particular, if γ is homotopic to a point in U, then

$$\int_\gamma f = 0.$$

Either of these statements may be viewed as a form of Cauchy's theorem. We prove Theorem 5.2 in detail, and leave Theorem 5.1 to the reader; the proof is entirely similar using Lemma 4.3 instead of Lemma 4.4 from the preceding section. The idea is that the homotopy gives us a finite sequence of paths close to each other in the sense of these lemmas, so that the integral of f over each successive path is unchanged.

The formal proof runs as follows. Let

$$\psi : [a, b] \times [c, d] \to U$$

be the homotopy. The image of ψ is compact, and hence has distance > 0 from the complement of U. By uniform continuity we can therefore find partitions

$$a = a_0 \leqq a_1 \leqq \cdots \leqq a_n = b,$$
$$c = c_0 \leqq c_1 \leqq \cdots \leqq c_m = d$$

of these intervals, such that if

$$S_{ij} = \text{small rectangle } [a_i, a_{i+1}] \times [b_j, b_{j+1}]$$

then the image $\psi(S_{ij})$ is contained in a disc D_{ij} which is itself contained in U. Let ψ_j be the continuous curve defined by

$$\psi_j(t) = \psi(t, c_j), \qquad j = 0, \ldots, m.$$

Then the continuous curves ψ_j, ψ_{j+1} are close together, and we can apply the lemma of the preceding section to conclude that

$$\int_{\psi_j} f = \int_{\psi_{j+1}} f.$$

Since $\psi_0 = \gamma$ and $\psi_m = \eta$, we see that the theorem is proved.

Remark. It is usually not difficult, although sometimes it is tedious, to exhibit a homotopy between continuous curves. Most of the time, one can achieve this homotopy by simple formulas when the curves are given explicitly.

Example. Let z, w be two points in the complex numbers. The segment between z, w, denoted by $[z, w]$, is the set of points

$$z + t(w - z), \qquad 0 \leqq t \leqq 1,$$

or equivalently,

$$(1 - t)z + tw, \qquad 0 \leqq t \leqq 1.$$

A set S of complex numbers is called **convex**, if, whenever $z, w \in S$, then the segment $[z, w]$ is also contained in S. We observe that a disc and a rectangle are convex.

Lemma 5.3. *Let S be a convex set, and let γ, η be continuous closed curves in S. Then γ, η are homotopic in S.*

Proof. We define

$$\psi(t, s) = s\gamma(t) + (1 - s)\eta(t).$$

It is immediately verified that each curve ψ_s defined by $\psi_s(t) = \psi(t, s)$ is a closed curve, and ψ is continuous. Also

$$\psi(t, 0) = \eta(t) \qquad \text{and} \qquad \psi(t, 1) = \gamma(t),$$

so the curves are homotopic. Note that the homotopy is given by a linear function, so if γ, η are smooth curves, that is C^1 curves, then each curve ψ_s is also of class C^1.

EXERCISE III §5

1. A set S is called **star-shaped** if there exists a point z_0 in S such that the line seg-
    ment between z_0 and any point z in S is contained in S. Prove that a star-
    shaped set is simply connected, that is, every closed path is homotopic to a
    point.

III §6. Existence of Global Primitives. Definition of the Logarithm

In §4 we constructed locally a primitive for a holomorphic function by integrating. We now have the means of constructing primitives for a
much wider class of open sets.

We say that an open set U is **simply connected** if it is connected and if
every closed path in U is homotopic to a point.

Theorem 6.1. *Let f be holomorphic on a simply connected open set U.
Let $z_0 \in U$. For any point $z \in U$ the integral*

$$g(z) = \int_{z_0}^{z} f(\zeta)\, d\zeta$$

*is independent of the path in U from z_0 to z, and g is a primitive for f,
namely $g'(z) = f(z)$.*

Proof. Let γ_1, γ_2 be two paths in U from z_0 to z. Let γ_2^- be the reverse
path of γ_2, from z to z_0. Then

$$\gamma = \{\gamma_1, \gamma_2^-\}$$

is a closed path, and by the first form of Cauchy's theorem,

$$\int_{\gamma_1} f + \int_{\gamma_2^-} f = \int_{\gamma} f = 0.$$

Since the integral of f over γ_2^- is the negative of the integral of f over γ_2, we have proved the first assertion.

As to the second, to prove the differentiability of g at a point z_1, if z is near z_1, then we may select a path from z_0 to z by passing through z_1, that is

$$g(z) = g(z_1) + \int_{z_1}^{z} f,$$

and we have already seen that this latter integral defines a local primitive for f in a neighborhood of z_1. Hence

$$g'(z) = f(z),$$

as desired.

Example. Let U be the plane from which a ray starting from the origin has been deleted. Then U is simply connected.

Proof. Let γ be any closed path in U. For simplicity, suppose the ray is the negative x-axis, as on Fig. 17. Then the path may be described in terms of polar coordinates,

$$\gamma(t) = r(t)e^{i\theta(t)}, \qquad a \le t \le b,$$

with $-\pi < \theta(t) < \pi$. We define the homotopy by

$$\psi(t, u) = r(ua + (1 - u)t)e^{i\theta(t)(1 - u)}, \qquad 0 \le u \le 1.$$

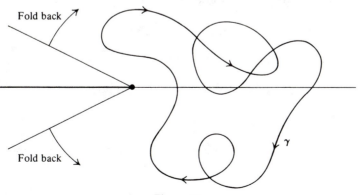

Figure 17

Geometrically, we are folding back the angle towards 0, and we are con-
tracting distance $r(t)$ towards $r(a)$. It is clear that ψ has the desired
property.

Example.   Definition of the logarithm.   Let U be a simply connected
open set not containing 0.   Pick a point $z_0 \in U$.   Let w_0 be a complex
number such that

$$e^{w_0} = z_0.$$

(Any two such numbers differ by an integral multiple of $2\pi i$.)   Define

$$\log z = w_0 + \int_{z_0}^{z} \frac{1}{\zeta} \, d\zeta.$$

Then $\log z$ (which depends on the choice of z_0 and w_0 only) is a primitive
for $1/z$ on U, and any other primitive differs from this one by a constant.
 If z is near z_0, then the function

$$w_0 + \log_0(1 + (z - z_0)/z_0),$$

where $\log_0$ is the usual power series for the logarithm near 1, defines a
analytic function near z_0, whose derivative is also $1/z$.   Since this function
and the above defined $\log z$ have the same value at z_0, namely w_0, they
must be equal near z_0.   Consequently we find that

$$e^{\log z} = z$$

for z near z_0.   The two analytic functions $e^{\log z}$ and z are equal near z_0.
Since U is connected, they are equal on U, and the above equation remains valid for all z in U.
 It is then easy to see that if $L(z)$ is a primitive for $1/z$ on U such that
$e^{L(z)} = z$, then there exists an integer k such that

$$L(z) = \log z + 2\pi i k.$$

We leave this to the reader.

 If U is a simply connected set containing a point

$$z_0 = r_0 e^{i\theta_0} \neq 0$$

with $-\pi < \theta_0 < \pi$, and $\log z$ is a determination of the logarithm on U
such that

$$\log z_0 = \log r_0 + i\theta_0,$$

and log r_0 is the usual real logarithm, then this is called the **principal value** of the logarithm.

Definition of z^α for any Complex α. By using the logarithm, we can define z under the following conditions.

Let U be simply connected not containing 0. Let α be a complex number $\neq 0$. Fix a determination of the log on U. With respect to this determination, we define

$$z^\alpha = e^{\alpha \, \log z}.$$

Then z^α is analytic on U.

Example. Let U be the open set obtained by deleting the positive real axis from the complex plane. We define the log to have the values

$$\log re^{i\theta} = \log r + i\theta,$$

where $0 < \theta < 2\pi$. This is also called a **principal value** for the log in that open set. Then

$$\log i = i\pi/2 \quad \text{and} \quad \log(-i) = 3\pi i/2.$$

In this case,

$$i^i = e^{i \, \log i} = e^{\pi i^2/2} = e^{-\pi/2}.$$

EXERCISES III §6

1. Compute the following values when the log is defined by its principal value on the open set U equal to the plane with the positive real axis deleted.
   (a) $\log i$ (b) $\log(-i)$ (c) $\log(-1 + i)$
   (d) i^i (e) $(-i)^i$ (f) $(-1)^i$
   (g) $(-1)^{-i}$ (h) $\log(-1 - i)$

2. Compute the values of the same expressions as in Exercise 1 (except (f) and (g)) when the open set consists of the plane from which the negative real axis has been deleted. Then take $-\pi < \theta < \pi$.

3. Let U be the plane with the negative real axis deleted. Find the limit

$$\lim_{y \to 0} [\log(a + iy) - \log(a - iy)]$$

where $a > 0$, and also where $a < 0$.

4. Let U be the plane with the positive real axis deleted. Find the limit

$$\lim_{y \to 0} [\log(a + iy) - \log(a - iy)]$$

where $a < 0$, and also where $a > 0$.

5. Over what kind of open sets could you define a holomorphic function $z^{1/3}$, or more generally $z^{1/n}$ for any positive integer n? Give examples, taking the open set to be as "large" as possible.

6. Let U be a simply connected open set. Let f be holomorphic on U and assume that $f(z) \neq 0$ for all $z \in U$.
   (a) Show how to define $\log(f(z))$ as a holomorphic function on U, by means of an integral.
   (b) Show that there exists a holomorphic function g on U such that $g^2 = f$. Does this last assertion remain true if 2 is replaced by an arbitrary positive integer n?

CHAPTER IV

Cauchy's Theorem, Second Part

We wish to give a general global criterion when the integral of a holomorphic function along a closed path is 0. In practice, we meet two types of properties of paths: (1) properties of homotopy, and (2) properties having to do with integration, relating to the number of times a curve "winds" around a point, as we already saw when we evaluated the integral

$$\int \frac{1}{\zeta - z} d\zeta$$

along a circle centered at z. These properties are of course related, but they also exist independently of each other, so we now consider those conditions on a closed path γ when

$$\int_{\gamma} f = 0$$

for all holomorphic functions f, and also describe what the value of this integral may be if not 0.

We shall give two proofs for the global version of Cauchy's theorem. Artin's proof depends only on Goursat's theorem for the integral of a holomorphic function around a rectangle, and a self-contained topological lemma, having only to do with paths and not holomorphic functions. Dixon's proof uses some of the applications to holomorphic functions which bypass the topological considerations.

In this chapter, paths are again assumed to be piecewise C^1, and curves are again C^1.

IV §1. The Winding Number

In an example of Chapter III, §2, we found that

$$\frac{1}{2\pi i} \int_\gamma \frac{1}{z} \, dz = 1,$$

if γ is a circle around the origin, oriented counterclockwise. It is therefore reasonable to define for any closed path γ its **winding number with respect to a point** α to be

$$W(\gamma, \alpha) = \frac{1}{2\pi i} \int_\gamma \frac{1}{z - \alpha} \, dz,$$

provided the path does not pass through α. If γ is a curve defined on an interval $[a, b]$, then this integral can be written in the form

$$\int_\gamma \frac{1}{z - \alpha} \, dz = \int_a^b \frac{\gamma'(t)}{\gamma(t) - \alpha} \, dt.$$

Intuitively, the integral of $1/(z - \alpha)$ should be called $\log(z - \alpha)$, but it depends on the path. Later, we shall analyze this situation more closely, but for the moment, we need only the definition above without dealing with the log formally, although the interpretation in terms of the log is suggestive.

The definition of the winding number would be improper if the following lemma were not true.

Lemma 1.1. *If γ is a closed path, then $W(\gamma, \alpha)$ is an integer.*

Proof. Let $\gamma = \{\gamma_1, \dots, \gamma_n\}$ where each γ_i is a curve defined on an interval $[a_i, b_i]$. After a reparametrization of each curve if necessary, cf. Exercise 5 of Chapter III, §2, we may assume without loss of generality that $b_i = a_{i+1}$ for $i = 1, \dots, n - 1$. Then γ is defined and continuous on an interval $[a, b]$, where $a = a_1$, $b = b_n$, and γ is differentiable on each open interval $]a_i, b_i[$, (at the end points, γ is merely right and left differentiable). Let

$$F(t) = \int_a^t \frac{\gamma'(t)}{\gamma(t) - \alpha} \, dt.$$

Then F is continuous on $[a, b]$ and differentiable for $t \neq a_i, b_i$. Its derivative is

$$F'(t) = \frac{\gamma'(t)}{\gamma(t) - \alpha}.$$

(Intuitively, $F(t) = \log(\gamma(t) - \alpha)$ except for the dependence of path and a constant of integration, but this suggests our next step.) We compute the derivative of another function:

$$\frac{d}{dt} e^{-F(t)}(\gamma(t) - \alpha) = e^{-F(t)}\gamma'(t) - F'(t)e^{-F(t)}(\gamma(t) - \alpha) = 0.$$

Hence there is a constant C such that $e^{-F(t)}(\gamma(t) - \alpha) = C$, so

$$\gamma(t) - \alpha = Ce^{F(t)}.$$

Since γ is a closed path, we have $\gamma(a) = \gamma(b)$, and

$$Ce^{F(b)} = \gamma(b) - \alpha = \gamma(a) - \alpha = Ce^{F(a)}.$$

Since $\gamma(a) - a \neq 0$ we conclude that $C \neq 0$, so that

$$e^{F(a)} = e^{F(b)}.$$

Hence there is an integer k such that

$$F(b) = F(a) + 2\pi i k.$$

But $F(a) = 0$, so $F(b) = 2\pi i k$, thereby proving the lemma.

The winding number of the curve in Fig. 1 with respect to α is equal to 2.

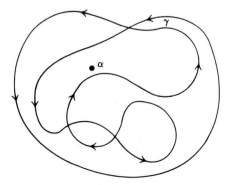

Figure 1

Lemma 1.2. *Let γ be a path. Then the function of α defined by*

$$\alpha \mapsto \int_\gamma \frac{1}{z - \alpha}\, dz$$

for α not on the path, is a continuous function of α.

Proof. Given α_0 not on the path, we have to see that

$$\int_\gamma \left(\frac{1}{z - \alpha} - \frac{1}{z - \alpha_0} \right) dz$$

tends to 0 as α tends to α_0. This integral is estimated as follows. The function $t \mapsto |\alpha_0 - \gamma(t)|$ is continuous and not 0, hence it has a minimum, the minimum distance between α_0 and the path, say

$$\min_t |\alpha_0 - \gamma(t)| = s.$$

If α is sufficiently close to α_0, then $|\alpha - \gamma(t)| \geq s/2$, as illustrated in Fig. 2. We have

$$\frac{1}{z - \alpha} - \frac{1}{z - \alpha_0} = \frac{\alpha - \alpha_0}{(z - \alpha)(z - \alpha_0)}$$

whence the estimate

$$\left| \frac{1}{z - \alpha} - \frac{1}{z - \alpha_0} \right| \leq \frac{1}{s^2/4} |\alpha - \alpha_0|.$$

Consequently, we get

$$\left| \int_\gamma \left(\frac{1}{z - \alpha} - \frac{1}{z - \alpha_0} \right) dz \right| \leq \frac{1}{s^2/4} |\alpha - \alpha_0| L(\gamma).$$

The right-hand side tends to 0 as α tends to α_0, and the continuity is proved.

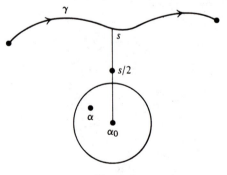

Figure 2

Lemma 1.3. *Let γ be a closed path. Let S be a connected set not intersecting γ. Then the function*

$$\alpha \mapsto \int_\gamma \frac{1}{z - \alpha}\, dz$$

is constant for α in S. If S is not bounded, then this constant is 0.

Proof. We know that from Lemma 1.1 that the integral is the winding number, and is therefore an integer. If a function takes its values in the integers, and is continuous, then it is constant on any curve, and consequently constant on a connected set. If S is not bounded, then for α arbitrarily large, the integrand has arbitrarily small absolute value, that is,

$$\frac{1}{|z - \alpha|}$$

is arbitrarily small, and estimating the integral shows that it must be equal to 0, as desired.

Example. Let U be the open set in Fig. 3. Then the set of points not in U consists of two connected components, one inside U and the other unbounded. Let γ be the closed curve shown in the figure, and let α_1 be the point inside γ, whereas α_2 is the point outside U, in the unbounded connected region. Then

$$W(\gamma, \alpha_1) = 1, \quad \text{but} \quad W(\gamma, \alpha_2) = 0.$$

We have drawn a curve extending from α_2 towards infinity, such that $W(\gamma, \alpha) = 0$ for α on this curve, according to the argument of Lemma 1.3.

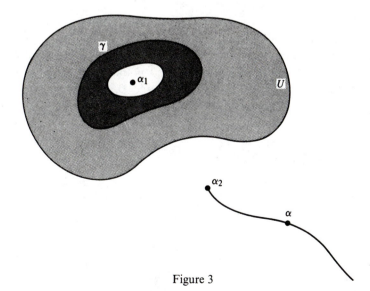

Figure 3

IV §2. Statement of Cauchy's Theorem

Let U be an open set. Let γ be a closed path in U. We want to give conditions that

$$\int_\gamma f = 0$$

for every holomorphic function f on U. We already know from the example of a winding circle that if the path winds around some point outside of U (in this example, the center of the circle), then definitely we can find functions whose integral is not equal to 0, and even with the special functions

$$f(z) = \frac{1}{z - \alpha},$$

where α is a point not in U. The remarkable fact about Cauchy's theorem is that it will tell us this is the only obstruction possible to having

$$\int_\gamma f = 0$$

for all possible functions f. In other words, the functions

$$\frac{1}{z - \alpha}, \qquad \alpha \notin U,$$

suffice to determine the behavior of $\int_\gamma f$ for all possible functions. With this in mind, we want to give a name to those closed paths in U having the property that they do not wind around points in the complement of U. The name we choose is homologous to 0, for historical reasons. Thus formally, we say that a closed path γ in U is **homologous to 0 in** U if

$$\int_\gamma \frac{1}{z - \alpha}\, dz = 0$$

for every point α not in U, or in other words, more briefly,

$$W(\gamma, \alpha) = 0$$

for every such point.

Similarly, let γ, η be closed paths in U. We say that they are **homologous in** U if

$$W(\gamma, \alpha) = W(\eta, \alpha)$$

for every point α in the complement of U. It will also follow from Cauchy's theorem that if γ and η are homologous, then

$$\int_\gamma f = \int_\eta f$$

for all holomorphic functions f on U.

Theorem 2.1

(i)   *If γ, η are closed paths in U and are homotopic, then they are homologous.*

(ii) *If γ, η are closed paths in U and are close together then they are homologous.*

Proof. The first statement follows from Theorem 5.2 of the preceding chapter because the function $1/(z - \alpha)$ is analytic on U for $\alpha \notin U$. The second statement is a special case of Lemma 4.4 of the preceding chapter.

Next we draw some examples of homologous paths.

In Fig. 4, the curves γ and η are **homologous**. Indeed, if α is a point inside the curves, then the winding number is 1, and if α is a point in the connected part going to infinity, then the winding number is 0.

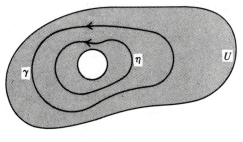

Figure 4

In Fig. 5 the path indicated is supposed to go around the top hole counterclockwise once, then around the bottom hole counterclockwise once, then around the top in the opposite direction, and then around the bottom in the opposite direction. This path is homologous to 0.

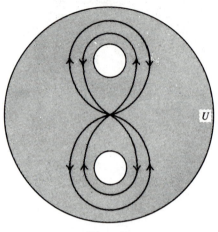

Figure 5

In Fig. 6, we are dealing with a simple closed curve, whose inside is contained in U, and the figure is intended to show that γ can be deformed to a point, so that γ is homologous to 0.

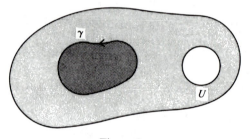

Figure 6

Given an open set U, we wish to determine in a simple way those closed paths which are not homologous to 0. For instance, the open set U might be as in Fig. 7, with three holes in it, at points z_1, z_2, z_3, so these points are assumed not to be in U.

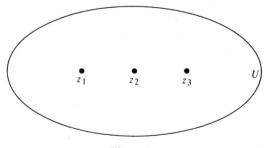

Figure 7

Let γ be a closed path in U, and let f be holomorphic on U. We illustrate γ in Fig. 8.

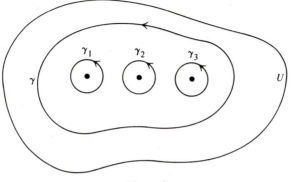

Figure 8

In that figure, we see that γ winds around the three points, and winds once. Let $\gamma_1, \gamma_2, \gamma_3$ be small circles centered at z_1, z_2, z_3 respectively, and oriented counterclockwise, as shown on Fig. 8. Then it is reasonable to expect that

$$\int_\gamma f = \int_{\gamma_1} f + \int_{\gamma_2} f + \int_{\gamma_3} f.$$

This will in fact be proved after Cauchy's theorem. We observe that taking $\gamma_1, \gamma_2, \gamma_3$ together does not constitute a "path" in the sense we have used that word, because, for instance, they form a disconnected set. However, it is convenient to have a terminology for a formal sum like $\gamma_1 + \gamma_2 + \gamma_3$, and to give it a name η, so that we can write

$$\int_\gamma f = \int_\eta f.$$

The name that is standard is the name **chain**. Thus let, in general, $\gamma_1, \ldots, \gamma_n$ be curves, and let $m_1, \ldots, m_n$ be integers which need not be positive. A formal sum

$$\gamma = m_1\gamma_1 + \cdots + m_n\gamma_n = \sum_{i=1}^n m_i\gamma_i$$

will be called a **chain**. If each curve γ_i is a curve in an open set U, we call γ a **chain in** U. We say that the chain is **closed** if it is a finite sum of closed paths. If γ is a chain as above, we define

$$\int_\gamma f = \sum m_i \int_{\gamma_i} f.$$

If $\gamma = \sum m_i \gamma_i$ is a closed chain, where each γ_i is a closed path, then its winding number with respect to a point α not on the chain is defined as before,

$$W(\gamma, \alpha) = \frac{1}{2\pi i} \int_\gamma \frac{1}{z - \alpha}\, dz.$$

If γ, η are closed chains in U, then we have

$$W(\gamma + \eta, \alpha) = W(\gamma, \alpha) + W(\eta, \alpha).$$

We say that γ is **homologous to** η in U, and write $\gamma \sim \eta$, if

$$W(\gamma, \alpha) = W(\eta, \alpha)$$

for every point $\alpha \notin U$. We say that γ is **homologous to 0 in** U and write

$$W(\gamma, \alpha) = 0$$

for every point $\alpha \notin U$.

Example. Let γ be the curve illustrated in Fig. 9, and let U be the plane from which three points z_1, z_2, z_3 have been deleted. Let γ_1, γ_2, γ_3 be small circles centered at z_1, z_2, z_3 respectively, oriented counterclockwise. Then it will be shown after Cauchy's theorem that

$$\gamma \sim \gamma_1 + 2\gamma_2 + 2\gamma_3,$$

so that for any function f holomorphic on U, we have

$$\int_\gamma f = \int_{\gamma_1} f + 2 \int_{\gamma_2} f + 2 \int_{\gamma_3} f.$$

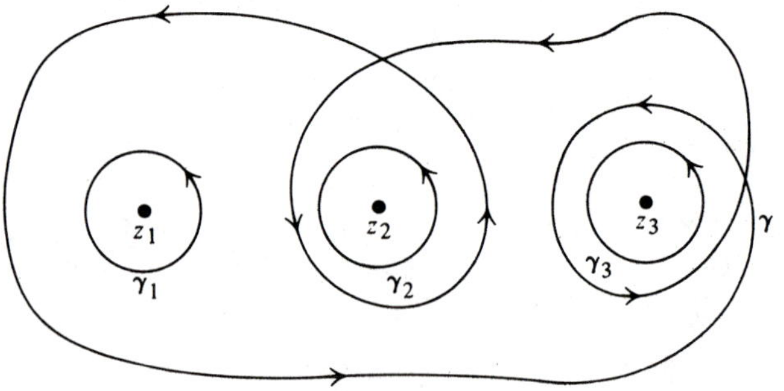

Figure 9

The above discussion and definition of chain provided motivation for what follows. We now go back to the formal development, and state Cauchy's theorem.

Theorem 2.2 (Cauchy's Theorem). *Let γ be a closed chain in an open set U, and assume that γ is homologous to 0 in U. Let f be holomorphic in U. Then*

$$\int_{\gamma} f = 0.$$

The proof will be given in the next sections. Observe that all we need of the holomorphic property is the existence of a primitive locally at every point of U, which was proved in the preceding chapter.

Corollary 2.3. *If γ, η are closed chains in U and γ, η are homologous in U, then*

$$\int_{\gamma} f = \int_{\eta} f.$$

Proof. Apply Cauchy's theorem to the closed chain $\gamma - \eta$.

Before giving the proof of Cauchy's theorem, we state two important applications, showing how one reduces integrals along complicated paths to integrals over small circles.

Theorem 2.4. *Let U be an open set and γ a closed chain in U such that γ is homologous to 0 in U. Let f be holomorphic on U except at a finite number of points $z_1,\ldots,z_n$. Let γ_i $(i = 1,\ldots,n)$ be the boundary of a closed disc D_i contained in U, containing z_i, and oriented counterclockwise. We assume that D_i does not intersect D_j if $i \neq j$. Let*

$$m_i = W(\gamma, z_i).$$

Let U^* be the set obtained by deleting $z_1,\ldots,z_n$ from U. Then γ is homologous to $\sum m_i\gamma_i$ in U^*, and in particular,

$$\int_{\gamma} f = \sum_{i=1}^{n} m_i \int_{\gamma_i} f.$$

Proof. Let $C = \gamma - \sum m_i\gamma_i$. Let α be a point outside U. Then

$$W(C, \alpha) = W(\gamma, \alpha) - \sum m_i W(\gamma_i, \alpha) = 0$$

because α is outside every small circle γ_i. If $\alpha = z_k$ for some k, then $W(\gamma_i, z_k) = 1$ if $i = k$ and 0 if $i \neq k$ by Lemma 1.3. Hence

$$W(C, z_k) = W(\gamma, z_k) - m_k = 0.$$

This proves that C is homologous to 0 in U^*. We apply Cauchy's theorem to conclude the proof.

The theorem is illustrated in Fig. 10. We have

$$\gamma \sim -\gamma_1 - 2\gamma_2 - \gamma_3 - 2\gamma_4,$$

and

$$\int_\gamma f = -\int_{\gamma_1} f - 2\int_{\gamma_2} f - \int_{\gamma_3} f - 2\int_{\gamma_4} f.$$

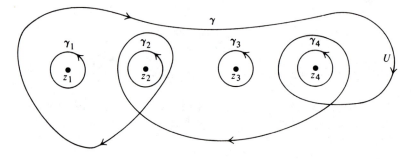

Figure 10

The theorem will be applied in many cases when U is a disc, say centered at the origin, and γ is a circle in U. Then certainly γ is homotopic to a point in U, and therefore homologous to 0 in U. Let $z_1, \ldots, z_n$ be points inside the circle, as on Fig. 11. Then Theorem 2.4 tells us that

$$\boxed{\int_\gamma f = \sum_{i=1}^n \int_{C_i} f,}$$

where C_i is a small circle around z_i. (Circles throughout are assumed oriented counterclockwise unless otherwise specified.)

In the next chapter, we shall give explicitly the values of the integrals around small circles in terms of the power series expansion of f around the points $z_1, \ldots, z_n$. We may also state the global version of **Cauchy's formula**.

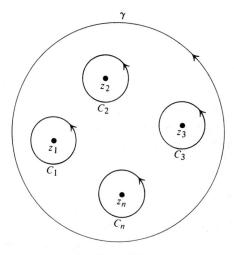

Figure 11

Theorem 2.5. *Let γ be a closed chain in U, homologous to 0 in U. Let f be holomorphic on U, let z_0 be in U and not on γ. Then*

$$\frac{1}{2\pi i} \int_\gamma \frac{f(z)}{z - z_0} \, dz = W(\gamma, z_0) f(z_0).$$

Proof. Let

$$g(z) = \frac{f(z) - f(z_0)}{z - z_0} \qquad \text{if} \quad z \neq z_0.$$

Since f is differentiable at z_0, we can define $g(z_0) = f'(z_0)$ so that g is also continuous at z_0, and in particular is bounded in a neighborhood of z_0. Let C_r be the circle of radius r centered at z_0. By Theorem 2.4 we have

$$\int_\gamma \frac{f(z) - f(z_0)}{z - z_0} \, dz = \int_\gamma g(z) \, dz = W(\gamma, z_0) \int_{C_r} g(z) \, dz.$$

Since g is bounded near z_0, and since the length of C_r tends to 0 as r tends to 0, we see that the right-hand side approaches 0 in absolute value as $r \to 0$. The left-hand side is independent of r, and is therefore equal to 0. Using the additivity of the integral yields

$$\int_\gamma \frac{f(z)}{z - z_0} \, dz = \int_\gamma \frac{f(z_0)}{z - z_0} \, dz,$$

whence the theorem follows.

Example.   Find the integral

$$\int_\gamma \frac{e^z}{z} \, dz$$

taken over the unit circle.   Here we let U be an open disc containing the closed unit disc.   Then γ is homologous to 0 in U, and in fact γ is homotopic to a point.   Hence Theorem 2.5 applies.   We let $z_0 = 0$.   The winding number of γ with respect to z_0 is 1, so

$$\int_\gamma \frac{e^z}{z} \, dz = 2\pi i e^0 = 2\pi i.$$

Remark 1.   We have shown that Theorem 2.4 (Cauchy's theorem) implies Theorem 2.5 (Cauchy's formula).   Conversely, it is easily seen that Cauchy's formula imples Cauchy's theorem.   Namely, we let z_0 be a point in U not on γ, and we let

$$F(z) = (z - z_0) f(z).$$

Applying Cauchy's formula to F yields

$$\frac{1}{2\pi i} \int_\gamma f(z) \, dz = \frac{1}{2\pi i} \int_\gamma \frac{F(z)}{z - z_0} \, dz = F(z_0) W(\gamma, z_0) = 0,$$

as desired.

Remark 2.   In older texts, Cauchy's theorem is usually stated for the integral over a simple closed curve, in the following form:

Let U be an open set, f holomorphic on U and let γ be a simple closed curve whose interior is contained in U.   Then

$$\int_\gamma f = 0.$$

It was realized for a long time that it is rather hard to prove that a simple closed curve decomposes the plane into two regions, its interior and exterior.   It is not even easy to define what is meant by "interior" or "exterior" a priori.   In fact, the theorem would be that the plane from which one deletes the curve consists of two connected sets.   For all points in one of the sets the winding number with respect to the curve is 1, and for all points in the other, the winding number is 0.   In any case, these general results are irrelevant in the applications.   Indeed, both in theoretical work and in practical applications, the statement of Cauchy's theorem as we gave it is quite efficient. In special cases, it is usually immediate to define the "interior" and "exterior" having the above property, for instance for circles or rectangles. One can apply Theorem 2.2 without appealing to any complicated result about general closed curves.

EXERCISES IV §2

1. Let f be holomorphic in an open set U. Let $z_0 \in U$ and let γ be a circle centered at z_0 such that the closed disc bounded by the circle is contained in U. Show that

$$\frac{1}{2\pi i} \int \frac{f(z)}{z - z_0} dz = f(z_0).$$

2. Find the integrals over the unit circle γ:

(a) $\displaystyle \int_\gamma \frac{\cos z}{z} dz$ (b) $\displaystyle \int_\gamma \frac{\sin z}{z} dz$ (c) $\displaystyle \int_\gamma \frac{\cos(z^2)}{z} dz$

3. (a) Show that the association $f \mapsto f'/f$ (where f is holomorphic) sends products to sums.
   (b) If $P(z) = (z - a_1) \cdots (z - a_n)$, where $a_1, \ldots, a_n$ are the roots, what is P'/P?
   (c) Let γ be a closed path such that none of the roots of P lie on γ. Show that

$$\frac{1}{2\pi i} \int_\gamma (P'/P)(z) \, dz = W(\gamma, z_1) + \cdots + W(\gamma, z_n).$$

IV §3. Artin's Proof

We have already found that integrating along sides of a rectangle works better than over arbitrary curves. We pursue this idea. A path will be said to be **rectangular** if every curve of the path is either a horizontal segment or a vertical segment. We shall see that every path is homologous with a rectangular path, and in fact we prove:

Lemma 3.1. *Let γ be a path in an open set U. Then there exists a rectangular path η with the same end points, and such that γ, η are close together in U in the sense of Chapter III, §4. In particular, γ and η are homologous in U, and for any holomorphic function f on U we have*

$$\int_\gamma f = \int_\eta f.$$

Proof. Suppose γ is defined on an interval $[a, b]$. We take a partition of the interval,

$$a = a_0 \leq a_1 \leq a_2 \leq \cdots \leq a_n = b$$

such that the image of each small interval

$$\gamma([a_i, a_{i+1}])$$

is contained in a disc D_i on which f has a primitive. Then we replace the curve γ on the interval $[a_i, a_{i+1}]$ by the rectangular curve drawn on Fig. 12. This proves the lemma.

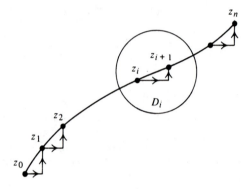

Figure 12

In the figure, we let $z_i = \gamma(a_i)$.

If γ is a closed path, then it is clear that the rectangular path constructed in the lemma is also a closed path, looking like this:

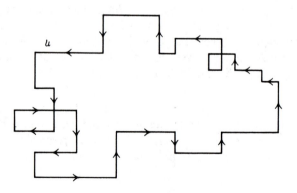

Figure 13

The lemma reduces the proof of Cauchy's theorem to the case when γ is a rectangular closed chain. We shall now reduce Cauchy's theorem to the case of rectangles by stating and proving a theorem having nothing to do with holomorphic functions. We need a little more terminology.

Let γ be a curve in an open set U, defined on an interval $[a, b]$. Let

$$a = a_0 \leqq a_1 \leqq a_2 \leqq \cdots \leqq a_n = b$$

be a partition of the interval. Let

$$\gamma_i \colon [a_i, a_{i+1}] \to U$$

be the restriction of γ to the smaller interval $[a_i, a_{i+1}]$. Then we agree to call the chain

$$\gamma_1 + \gamma_2 + \cdots + \gamma_n$$

a **subdivision** of γ. Furthermore, if η_i is obtained from γ_i by another parametrization, we again agree to call the chain

$$\eta_1 + \eta_2 + \cdots + \eta_n$$

a **subdivision** of γ. For any practical purposes, the chains γ and

$$\eta_1 + \eta_2 + \cdots + \eta_n$$

do not differ from each other. In Fig. 14 we illustrate such a chain γ and a subdivision $\eta_1 + \eta_2 + \eta_3 + \eta_4$.

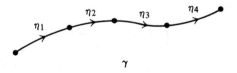

Figure 14

Similarly, if $\gamma = \sum m_i \gamma_i$ is a chain, and $\{\eta_{ij}\}$ is a subdivision of γ_i, we call

$$\sum_i \sum_j m_i \eta_{ij}$$

a **subdivision** of γ.

Theorem 3.2. *Let γ be a rectangular closed chain in U, and assume that γ is homologous to 0 in U, i.e.*

$$W(\gamma, \alpha) = 0$$

for every point α not in U. Then there exist rectangles $R_1, \ldots, R_N$ contained in U, such that if ∂R_i is the boundary of R_i oriented counterclockwise, then a subdivision of γ is equal to

$$\sum_{i=1}^{N} m_i \cdot \partial R_i$$

for some integers m_i.

Theorem 3.2 makes Cauchy's theorem obvious because we know that for any holomorphic function f on U, we have

$$\int_{\partial R_i} f = 0$$

by Goursat's theorem. Hence the integral of f over the subdivision of γ is also equal to 0, whence the integral of f over γ is also equal to 0.

We now prove the thorem. Given the rectangular chain γ, we draw all vertical and horizontal lines passing through the sides of the chain, as illustrated on Fig. 15.

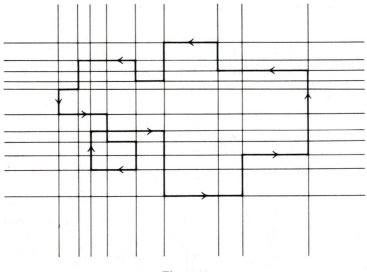

Figure 15

Then these vertical and horizontal lines decompose the plane into rectangles, and rectangular regions extending to infinity in the vertical and horizontal direction. Let R_i be one of the rectangles, and let $α_i$ be a point inside R_i. Let

$$m_i = W(γ, α_i).$$

For some rectangles we have $m_i = 0$, and for some rectangles, we have $m_i \neq 0$. We let $R_1,\ldots,R_N$ be those rectangles such that $m_1,\ldots,m_N$ are not 0, and we let ∂R_i be the boundary of R_i for $i = 1,\ldots,N$, oriented counterclockwise. We shall prove:

1. *Every rectangle R_i such that $m_i \neq 0$ is contained in U.*

2. *Some subdivision of γ is equal to*

$$\sum_{i=1}^{N} m_i \, \partial R_i.$$

This will prove the desired theorem.

Assertion 1. By assumption, α_i must be in U, because $W(\gamma, \alpha) = 0$ for every point α outside of U. Since the winding number is constant on connected sets, it is constant on the interior of R_i, hence $\neq 0$, and the interior of R_i is contained in U. If a boundary point of R_i is on γ, then it is in U. If a boundary point of R_i is not on γ, then the winding number with respect to γ is defined, and is equal to $m_i \neq 0$ by continuity (Lemma 1.2). This proves that the whole rectangle R_i, including its boundary, is contained in U, and proves the first assertion.

Assertion 2. We now replace γ by an appropriate subdivision. The vertical and horizontal lines cut γ in various points. We can then find a subdivision η of γ such that every curve occurring in η is some side of a rectangle, or the finite side of one of the infinite rectangular regions. The subdivision η is the sum of such sides, taken with appropriate multiplicities. It will now suffice to prove that

$$\eta = \sum m_i \, \partial R_i.$$

Let C be the closed chain

$$C = \eta - \sum_{i=1}^{N} m_i \, \partial R_i.$$

Then for every point α not on C we have

$$W(C, \alpha) = 0.$$

Proof. If α lies in one of the infinite regions, then $W(C, \alpha) = 0$ by Lemma 1.3. Suppose that α lies in some rectangle R. If R is one of the rectangles R_i, say $R = R_k$, then

$$W(\partial R_k, \alpha) = 1 \qquad \text{and} \qquad W(\partial R_i, \alpha) = 0 \qquad \text{if} \quad i \neq k,$$

because α is outside R_i if $i \neq k$ and Lemma 1.3 applies again. Then

$$W(C, \alpha) = W(\gamma, \alpha) - m_k W(\partial R_k, \alpha) = 0$$

by definition of m_k. On the other hand, if R is not one of the rectangles R_i, then $W(\gamma, \alpha) = 0$ by definition, and also

$$W(\partial R_i, \alpha) = 0 \qquad \text{for all} \quad i = 1, \dots, N$$

by Lemma 1.3. Thus in this case also we get $W(C, \alpha) = 0$. If α lies on the boundary of a rectangle or of an infinite rectangular region, but not on C, then $W(C, \alpha)$ is 0 by the continuity of the winding number. This proves our assertion.

We now prove that $C = 0$. Suppose that $C \neq 0$, so that we can write

$$C = m\sigma + C^*,$$

where σ is a horizontal or vertical segment, m is an integer $\neq 0$, and C^* is a chain of vertical and horizontal segments other than σ. We have to distinguish two cases, according as σ is the side of some rectangle, or σ is the side of two infinite rectangular regions, illustrated in Fig. 16(a) and (b).

(a)   (b)

Figure 16

In the first case, we take σ with the orientation arising from the counterclockwise orientation of the boundary of the rectangle, and we denote the rectangle by R. Then the chain

$$C - m \cdot \partial R$$

does not contain σ. Let α be a point inside R, and let α' be a point near σ but on the opposite side from α, also as shown. Then we can join α to α' by a segment which does not intersect $C - m \cdot \partial R$. By continuity and the connectedness of the segment, we conclude that

$$W(C - m \cdot \partial R, \alpha) = W(C - m \cdot \partial R, \alpha').$$

But $W(m \cdot \partial R, \alpha) = m$ and $W(m \cdot \partial R, \alpha') = 0$ because α' is outside of R. Since $W(C, \alpha) = W(C, \alpha') = 0$, we conclude that $m = 0$, as desired.

Consider now the second case, illustrated in Fig. 16(b), when the two regions adjacent to σ are both infinite. We show that this cannot happen. In this case, we have

$$W(C, \alpha) = W(C, \alpha') = 0$$

by the usual Lemma 1.3, and therefore

$$m \int_\sigma \frac{1}{z - \alpha} \, dz + \int_{C^*} \frac{1}{z - \alpha} \, dz = m \int_\sigma \frac{1}{z - \alpha'} \, dz + \int_{C^*} \frac{1}{z - \alpha'} \, dz.$$

Subtracting, we obtain

$$m \int_\sigma \left(\frac{1}{z - \alpha} - \frac{1}{z - \alpha'} \right) dz = \int_{C^*} \left(\frac{1}{z - \alpha'} - \frac{1}{z - \alpha} \right) dz.$$

Since C^* does not contain σ, the right-hand side is a continuous function of α, α' when α, α' approach each other on a segment passing through σ. On the other hand, a direct integration on the left-hand side shows that the integral over σ is equal to a fixed non-zero multiple of $2\pi i$, and this is a contradiction unless $m = 0$.

We leave the direct integration to the reader. For instance, in case σ is a vertical segment, we parametrize it by picking some point, say z_0 in the middle, and letting the segment be

$$z_0 + it, \qquad -c \leq t \leq c,$$

for some $c > 0$. We pick $\alpha = z_0 + x$ and $\alpha' = z_0 - x$ for small values of $x > 0$. The integral on the left-hand side can be evaluated (it turns out to be an arctangent), and the limit as $x \to 0$ can be found by freshman calculus.

This concludes the proof that $C = 0$, whence a subdivision of γ is equal to

$$\sum m_i \cdot \partial R_i,$$

and Theorem 3.2 is proved.

CHAPTER V

Applications of Cauchy's Integral Formula

In the present chapter we show how a holomorphic function and the derivative of a holomorphic function can be expressed as an integral in terms of the function. We then give applications of this, getting the expansion of a holomorphic function into a power series, and studying the possible singularities which may arise when a function is analytic near a point, but may not be holomorphic at the point itself.

We shall be integrating along circles a lot. It is understood that a **circle**, unless otherwise specified, is always oriented counterclockwise.

V §1. Cauchy's Integral Formula on a Disc

In this section we shall apply Cauchy's integral formula by integrating over a circle, so we restate the theorem in this simple form.

Theorem 1.1. *Let f be holomorphic on the closed disc of radius R centered at a point, and let D be the open disc. Let γ be the circle which is the boundary of the disc. Then for any point z_0 in D we have*

$$f(z_0) = \frac{1}{2\pi i} \int_\gamma \frac{f(z)}{z - z_0} \, dz.$$

Next we see that a holomorphic function is analytic, and derive its power series expansion from the formula for its derivatives.

Theorem 1.2. *Let γ be a path in an open set U and let g be a continuous function on γ (i.e. on the image $\gamma([a, b])$ if γ is defined on $[a, b]$). If z is not on γ, define*

$$f(z) = \int_\gamma \frac{g(\zeta)}{\zeta - z} \, d\zeta.$$

Then f is analytic on the complement of γ in U, and its derivatives are given by

$$f^{(n)}(z) = n! \int_\gamma \frac{g(\zeta)}{(\zeta - z)^{n+1}} \, d\zeta.$$

Proof. Let $z_0 \in U$ and z_0 not on γ. Then there is some $r > 0$ such that $|\zeta - z_0| \geq r$ for all points ζ on γ. Let $0 < s < r$. Let $D(z_0, s)$ be the disc of radius s. We shall see that f has a power series expansion on this disc. We write

$$\frac{1}{\zeta - z} = \frac{1}{\zeta - z_0 - (z - z_0)} = \frac{1}{\zeta - z_0} \left(\frac{1}{1 - \dfrac{z - z_0}{\zeta - z_0}} \right)$$

$$= \frac{1}{\zeta - z_0} \left(1 + \frac{z - z_0}{\zeta - z_0} + \left(\frac{z - z_0}{\zeta - z_0} \right)^2 + \cdots \right).$$

This geometric series converges absolutely and uniformly for $|z - z_0| \leq s$ because

$$\left| \frac{z - z_0}{\zeta - z_0} \right| \leq s/r < 1.$$

The function g is bounded on γ. By Theorem 2.4 of Chapter III, we can therefore integrate term by term, and we find

$$f(z) = \sum_{n=0}^{\infty} \int_\gamma \frac{g(\zeta)}{(\zeta - z_0)^{n+1}} \, d\zeta \cdot (z - z_0)^n$$

$$= \sum_{n=0}^{\infty} a_n (z - z_0)^n,$$

where

$$a_n = \int_\gamma \frac{g(\zeta)}{(\zeta - z_0)^{n+1}} \, d\zeta.$$

This proves first that f is analytic, and gives us the coefficients of its power series expansion near z_0. We then know the derivatives of f at z_0 by Chapter II, §7, and thus conclude the proof of the present theorem.

In particular, we now see that **a function is analytic if and only if it is holomorphic.** The two words will be used interchangeably from now on.

The next theorem is an immediate consequence of the preceding ones, but is stated separately in view of its importance. It gives an indication of the maximal radius of convergence for the power series expansion of an analytic function.

Theorem 1.3. *Let f be analytic on a closed disc of radius R centered at z_0, and let*

$$f(z) = \sum a_n(z - z_0)^n$$

be the power series for f at z_0. Then

$$a_n = \frac{1}{n!} f^{(n)}(z_0) = \frac{1}{2\pi i} \int_{C_R} \frac{f(\zeta)}{(\zeta - z_0)^{n+1}} \, d\zeta,$$

where C_R is the circle of radius R centered at z_0 and if $\|f\|_R$ denotes the sup norm of f on this circle, then

$$|a_n| \leqq \|f\|_R / R^n.$$

In particular, the radius of convergence of the series is $\geqq R$.

Proof. Obvious as a special case of the preceding theorems. The estimate is taken as usual, equal to the product of the sup norm of the expression under the integral sign, and the length of the curve which is $2\pi R$. For all ζ on the circle, we have

$$|\zeta - z_0| = R,$$

so the desired estimate falls out. Taking the n-th root of $|a_n|$, we conclude at once that the radius of convergence is at least R.

A function f is called **entire** if it is holomorphic on all of $\mathbf{C}$.

Theorem 1.4 (Liouville's Theorem). *A bounded entire function is constant.*

Proof. If f is bounded, then $\|f\|_R$ is bounded for all R. In the preceding theorem, we let R tend to infinity, and conclude that the coefficients are all equal to 0 if $n \geqq 1$. This proves Liouville's theorem.

We have already proved that a polynomial always has a root in the complex numbers. We give here the more usual proof as a corollary of Liouville's theorem.

Let $f(z)$ be a non-constant polynomial,

$$f(z) = a_n z^n + \cdots + a_0,$$

with $a_n \neq 0$. Suppose that $f(z) \neq 0$ for all z. Then the function

$$g(z) = 1/f(z)$$

is defined for all z and analytic on $\mathbf{C}$. On the other hand, writing

$$f(z) = a_n z^n (1 + b_1/z + \cdots + b_n/z^n)$$

with appropriate constants $b_1, \ldots, b_n$ we see that $|f(z)|$ is large when $|z|$ is large, and hence that $|g(z)| \to 0$ as $|z| \to \infty$. For sufficiently large radius R, $|g(z)|$ is small for z outside the closed disc of radius R, and $|g(z)|$ has a maximum on this disc since the disc is compact. Hence g is a bounded entire function, and therefore constant by Liouville's theorem. This is obviously a contradiction, proving that f must have a zero somewhere in $\mathbf{C}$.

Theorem 1.5. *Let $\{f_n\}$ be a sequence of holomorphic functions on an open set U. Assume that for each compact subset K of U the sequence converges uniformly on K, and let the limit function be f. Then f is holomorphic.*

Proof. Let $z_0 \in U$, and let D_R be a closed disc of radius R centered at z_0 and contained in U. Then the sequence $\{f_n\}$ converges uniformly on D_R. Let C_R be the circle which is the boundary of D_R. Let $D_{R/2}$ be the closed disc of radius $R/2$ centered at z_0. Then for $z \in D_{R/2}$ we have

$$f_n(z) = \frac{1}{2\pi i} \int_{C_R} \frac{f_n(\zeta)}{\zeta - z} \, d\zeta,$$

and $|\zeta - z| \geq R/2$. Since $\{f_n\}$ converges uniformly, for $|z - z_0| \leq R/2$, we get

$$f(z) = \frac{1}{2\pi i} \int_{C_R} \frac{f(\zeta)}{\zeta - z} \, d\zeta.$$

By Theorem 1.2 it follows that f is holomorphic on a neighborhood of z_0. Since this is true for every z_0 in U, we have proved what we wanted.

Theorem 1.6. *Let $\{f_n\}$ be a sequence of analytic functions on an open set U, converging uniformly on every compact subset K of U to a function f. Then the sequence of derivatives $\{f_n'\}$ converges uniformly on every compact subset K, and $\lim f_n' = f'$.*

Proof. The proof will be left as an exercise to the reader. [Hint: Cover the compact set with a finite number of closed discs contained in U, and of sufficiently small radius. Cauchy's formula expresses the derivative f'_n as an integral, and one can argue as in the previous theorem.]

Example. Let

$$f(z) = \sum_{n=1}^{\infty} \frac{1}{n^z}.$$

We shall prove that this function is holomorphic for Re $z > 1$. Each term

$$f_n(z) = n^{-z} = e^{-z \log n}$$

is an entire function. Let $z = x + iy$. We have

$$|e^{-z \log n}| = |e^{-x \log n} e^{-iy \log n}| = n^{-x}.$$

Let $c > 1$. For $x \geq c$ we have $|n^{-z}| \leq n^{-c}$ and the series

$$\sum_{n=1}^{\infty} \frac{1}{n^c}$$

converges for $c > 1$. Hence the series $\sum f_n(z)$ converges uniformly and absolutely for Re $z \geq c$, and therefore defines a holomorphic function for Re $z > c$. This is true for every $c > 1$, and hence f is holomorphic for Re $z > 1$.

In the same example, we have

$$f'_n(z) = \frac{-\log n}{n^z}.$$

By Theorem 1.6, it follows that

$$f'(z) = \sum_{n=1}^{\infty} \frac{-\log n}{n^z}$$

in this same region.

EXERCISES V §1

1. Let f be analytic on an open set U, let $z_0 \in U$ and $f'(z_0) \neq 0$. Show that

$$\frac{2\pi i}{f'(z_0)} = \int_c \frac{1}{f(z) - f(z_0)} \, dz,$$

where C is a small circle centered at z_0.

2. Weierstrass' theorem for a real interval $[a, b]$ states that a continuous function can be uniformly approximated by polynomials. Is this conclusion still true for the closed unit disc, i.e. can every continuous function on the disc be uniformly approximated by polynomials?

3. Let $a > 0$. Show that each of the following series represents a holomorphic function:

(a) $\sum\limits_{n=1}^{\infty} e^{-an^2z}$ for $\mathrm{Re}\, z > 0$;

(b) $\sum\limits_{n=1}^{\infty} \dfrac{e^{-anz}}{(a+n)^2}$ for $\mathrm{Re}\, z > 0$;

(c) $\sum\limits_{n=1}^{\infty} \dfrac{1}{(a+n)^z}$ for $\mathrm{Re}\, z > 1$.

4. Show that each of the two series converges uniformly on each closed disc $|z| \leq c$ with $0 < c < 1$, and also prove the equality:

$$\sum_{n=1}^{\infty} \frac{nz^n}{1-z^n} = \sum_{n=1}^{\infty} \frac{z^n}{(1-z^n)^2}.$$

[*Hint*: For the equality, write each side in a double series, and reverse the order of summation.]

5. Let f be an entire function. Assume that there exist numbers $C > 0$ and $k > 0$ such that

$$\|f\|_R \leq CR^k$$

for all R sufficiently large. Show that f is a polynomial of degree $\leq k$.

6. **Dirichlet Series.** Let $\{a_n\}$ be a sequence of complex numbers. Show that the series $\sum a_n/n^s$, if it converges at all for some complex s, converges absolutely in a right half plane $\mathrm{Re}(s) > \sigma_0$, and uniformly in $\mathrm{Re}(s) > \sigma_0 + \epsilon$ for every $\epsilon > 0$. Show that the series defines an analytic function in this half plane. The number σ_0 is called the **abscissa of convergence**.

The next exercises give expressions and estimates for an analytic function in terms of integrals.

7. Let f be analytic on the closed unit disc D. Show that

$$\iint_D f(x+iy)\, dy\, dx = \pi f(0).$$

[*Hint*: Use polar coordinates and Cauchy's formula.]

For the next exercise, recall that a norm $\|\ \|$ on a space of functions associates to each function f a real number ≥ 0, satisfying the following conditions:

N 1.   We have $\|f\| = 0$ if and only if $f = 0$.

N 2.   If c is a complex number, then $\|cf\| = |c|\,\|f\|$.

N 3. $\|f + g\| \leq \|f\| + \|g\|$.

*8. Let A be the closure of a bounded open set in the plane. Let f, g be continuous functions on A. Define their scalar product

$$\langle f, g \rangle = \int\int_A f(z)\,\overline{g(z)}\,dy\,dx$$

and define the associated L^2-**norm** by its square,

$$\|f\|_2^2 = \int\int_A |f(z)|^2\,dy\,dx.$$

Show that $\|f\|_2$ does define a norm. Define

$$\|f\|_1 = \int\int_A |f(z)|\,dy\,dx.$$

Show that $f \mapsto \|f\|_1$ is a norm on the space of continuous functions on A, called the L^1-**norm**. This is just preliminary. Prove:

(a) There exist constants $C_1, C_2 > 0$ such that, if f is analytic on a disc of radius R, centered at the origin, and $0 < s < R$, then

$$\|f\| \leqq C_1\|f\|_1 \leqq C_2\|f\|_2,$$

where $\|\ \|$ is the sup norm on the closed disc of radius s, and the L^1, L^2 norms also refer to the closed disc of radius s.

(b) Let $\{f_n\}$ be a sequence of holomorphic functions on an open set U, and assume that this sequence is L^2-Cauchy. Show that it converges uniformly on compact subsets of U.

*9. Let U, V be open discs centered at the origin. Let $f = f(z, w)$ be a *continuous* function on the product $U \times V$, such that for each w the function $z \mapsto f(z, w)$ and for each z the function $w \mapsto f(z, w)$ are analytic on U and V, respectively. Show that f has a power series expansion

$$f(z, w) = \sum a_{mn} z^m w^n$$

which converges absolutely and uniformly for $|z| \leqq r$ and $|w| \leqq r$, for some positive number r. [*Hint:* Apply Cauchy's formula for derivatives twice, with respect to the two variables to get an estimate for the coefficients a_{mn}.] Generalize to several variables instead of two variables.

Note. This exercise is really quite trivial, although it is not generally realized that it is so. *The point is that the function f is assumed to be continuous.* If that assumption is not made, the situation becomes much more difficult to handle, and the result is known as Hartogs' theorem. In practice, continuity is indeed satisfied.

V §2. Laurent Series

By a **Laurent series**, we mean a series

$$f(z) = \sum_{n=-\infty}^{\infty} a_n z^n.$$

Let A be a set of complex numbers. We say that the Laurent series **converges absolutely** (resp. uniformly) on A if the two series

$$f^+(z) = \sum_{n \geq 0} a_n z^n \qquad \text{and} \qquad f^-(z) = \sum_{n < 0} a_n z^n$$

converge absolutely (resp. uniformly) on A. If that is the case, then $f(z)$ is regarded as the sum,

$$f(z) = f^+(z) + f^-(z).$$

Let r, R be positive numbers with $0 \leq r < R$. We shall consider the annulus A consisting of all complex numbers z such that

$$r \leq |z| \leq R.$$

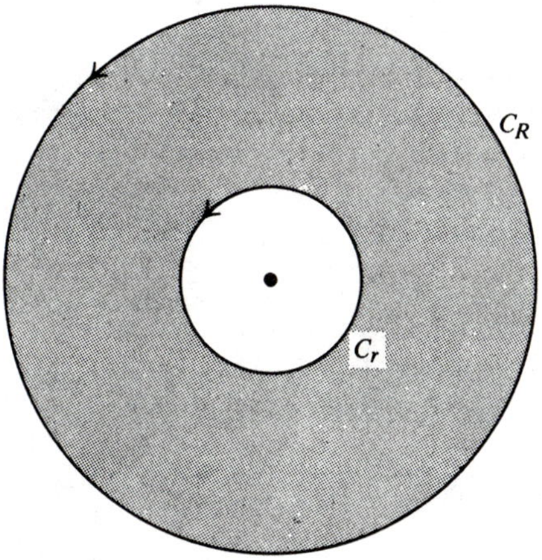

Figure 1

Theorem 2.1. *Let A be the above annulus, and let f be a holomorphic function on A. Let $r < s < S < R$. Then f has a Laurent expansion*

$$f(z) = \sum_{n=-\infty}^{\infty} a_n z^n$$

which converges absolutely and uniformly on $s \leq |z| \leq S$. *Let* C_R *and* C_r *be the circles of radius* R *and* r, *respectively. Then the coefficients* a_n *are obtained by the usual formula:*

$$a_n = \frac{1}{2\pi i} \int_{C_R} \frac{f(\zeta)}{\zeta^{n+1}} \, d\zeta \qquad \text{if} \quad n \geq 0,$$

$$a_n = \frac{1}{2\pi i} \int_{C_r} \frac{f(\zeta)}{\zeta^{n+1}} \, d\zeta \qquad \text{if} \quad n < 0.$$

Proof. For some $\epsilon > 0$ we may assume (by the definition of what it means for f to be holomorphic on the closed annulus) that f is holomorphic on the open annulus U of complex numbers z such that

$$r - \epsilon < |z| < R + \epsilon.$$

The chain $C_R - C_r$ is homologous to 0 on U, because if a point lies in the outer part then its winding number is zero by the usual Lemma 1.3 of Chapter IV, and if the point lies in the disc inside the annulus, then its winding number is 0. Cauchy's formula then implies that for z in the annulus,

$$f(z) = \frac{1}{2\pi i} \int_{C_R} \frac{f(\zeta)}{\zeta - z} \, d\zeta - \frac{1}{2\pi i} \int_{C_r} \frac{f(\zeta)}{\zeta - z} \, d\zeta.$$

We may now prove the theorem. The first integral is handled just as in the ordinary case of the derivation of Cauchy's formula, and the second is handled in a similar manner as follows. We write

$$\xi - z = -z\left(1 - \frac{\zeta}{z}\right).$$

Then

$$\left|\frac{\zeta}{z}\right| \leq r/s < 1,$$

so the geometric series converges,

$$\frac{1}{z}\frac{1}{1 - \zeta/z} = \frac{1}{z}\left(1 + \frac{\zeta}{z} + \left(\frac{\zeta}{z}\right)^2 + \cdots\right).$$

We can then integrate term by term, and the desired expansion falls out. We leave the uniqueness of the coefficients to the reader.

An example of a function with a Laurent series with infinitely many negative terms is given by $e^{1/z}$, that is, by substituting $1/z$ in the ordinary exponential series.

If an annulus is centered at a point z_0, then one obtains a Laurent series at z_0 of the form

$$f(z) = \sum_{n=-\infty}^{\infty} a_n(z - z_0)^n.$$

Example. We want to find the Laurent series for

$$f(z) = \frac{1}{z(z - 1)}$$

for $0 < |z| < 1$. We write f in partial fractions:

$$f(z) = \frac{1}{z - 1} - \frac{1}{z}.$$

Then for one term we get the geometric series,

$$\frac{1}{z - 1} = -\frac{1}{1 - z} = -(1 + z + z^2 + \cdots)$$

whence

$$f(z) = -\frac{1}{z} - 1 - z - z^2 - \cdots.$$

On the other hand, suppose we want the Laurent series for $|z| > 1$. Then we write

$$\frac{1}{z - 1} = \frac{1}{z}\left(\frac{1}{1 - 1/z}\right) = \frac{1}{z}\left(1 + \frac{1}{z} + \frac{1}{z^2} + \cdots\right)$$

whence

$$f(z) = \frac{1}{z^2} + \frac{1}{z^3} + \frac{1}{z^4} + \cdots.$$

EXERCISES V §2

1. Prove that the Laurent series can be differentiated term by term in the usual manner to give the derivative of f on the annulus.

2. Let f be holomorphic on the annulus A, defined by

$$0 < r \leq |z| \leq R.$$

Prove that there exist functions f_1, f_2 such that f_1 is holomorphic for $|z| \leq r$, f_2 is holomorphic for $|z| \geq r$ and

$$f = f_1 + f_2$$

on the annulus.

3. Let f be analytic on the whole complex plane, and assume that there exists $C > 0$ and an integer $N > 0$ such that for all $|z|$ sufficiently large we have

$$|f(z)| \leq C|z|^N.$$

Prove that f is a polynomial of degree $\leq N$.

4. Expand the function

$$f(z) = \frac{z}{1 + z^3}$$

(a) in a series of positive powers of z, and
(b) in a series of negative powers of z.
In each case, specify the region in which the expansion is valid.

5. Give the Laurent expansions for the following functions:
   (a) $z/(z + 2)$ for $|z| > 2$ (b) $\sin 1/z$ for $z \neq 0$
   (c) $\cos 1/z$ for $z \neq 0$ (d) $\dfrac{1}{(z - 3)}$ for $|z| > 3$

6. Prove the following expansions:

   (a) $e^z = e + e \sum\limits_{n=1}^{\infty} \dfrac{1}{n!}(z - 1)^n$

   (b) $1/z = \sum\limits_{n=0}^{\infty} (-1)^n(z - 1)^n$ for $|z - 1| < 1$

   (c) $1/z^2 = 1 + \sum\limits_{n=1}^{\infty} (n + 1)(z + 1)^n$ for $|z + 1| < 1$

7. Expand (a) $\cos z$, (b) $\sin z$ in a power series about $\pi/2$.

8. Let $f(z) = \dfrac{1}{(z - 1)(z - 2)}$. Find the Laurent series for f:
   (a) In the disc $|z| < 1$.
   (b) In the annulus $1 < |z| < 2$.
   (c) In the region $2 < |z|$.

9. Find the Laurent series for $(z + 1)/(z - 1)$ in the region (a) $|z| < 1$; (b) $|z| > 1$.

10. Find the Laurent series for $1/z^2(1 - z)$ in the regions:
    (a) $0 < |z| < 1$; (b) $|z| > 1$.

11. Find the power series expansion of

$$f(z) = \frac{1}{1 + z^2}$$

around the point $z = 1$, and find the radius of convergence of this series.

12. Find the Laurent expansion of

$$f(z) = \frac{1}{(z - 1)^2(z + 1)^2}$$

for $1 < |z| < 2$.

13. Obtain the first four terms of the Laurent series expansion of

$$f(z) = \frac{e^z}{z(z^2 + 1)}$$

valid for $0 < |z| < 1$.

*14. Assume that f is analytic in the upper half plane, and that f is periodic of period 1. Show that f has an expansion of the form

$$f = \sum_{-\infty}^{\infty} c_n e^{2\pi i n z},$$

where

$$c_n = \int_0^1 f(x + iy)e^{-2\pi i n(x + iy)} dx,$$

for any value of $y > 0$. [*Hint*: Show that there is an analytic function f^* on a disc from which the origin is deleted such that

$$f^*(e^{2\pi i z}) = f(z).$$

What is the Laurent series for f^*? Abbreviate $q = e^{2\pi i z}$.

*15. Assumptions being as in Exercise 13, suppose in addition that there exists $y_0 > 0$ such that $f(z) = f(x + iy)$ is bounded in the domain $y \geq y_0$. Prove that the coefficients c_n are equal to 0 for $n < 0$. Is the converse true? Proof?

V §3. Isolated Singularities

Let z_0 be a complex number and let D be an open disc centered at z_0. Let U be the open set obtained by removing z_0 from D. A function f which is analytic on U is said to have an **isolated singularity** at z_0. We suppose this is the case.

Removable Singularities

Theorem 3.1. *If f is bounded in some neighborhood of z_0, then one can define $f(z_0)$ in a unique way such that the function is also analytic at z_0.*

Proof. Say $z_0 = 0$. By §2, we know that f has a Laurent expansion

$$f(z) = \sum_{n \geq 0} a_n z^n + \sum_{n < 0} a_n z^n$$

for $0 < |z| < R$. We have to show $a_n = 0$ if $n < 0$. Let $n = -m$ with $m > 0$. We have

$$a_{-m} = \frac{1}{2\pi i} \int_{C_r} f(\zeta)\zeta^{m-1} \, d\zeta,$$

for any circle C_r of small radius r. Since f is assumed bounded near 0 it follows that the right-hand side tends to 0 as r tends to 0, whence $a_{-m} = 0$, as was to be shown. (The uniqueness is clear by continuity.)

In the case of Theorem 3.1 it is customary to say that z_0 is a **removable singularity**.

Poles

Suppose the Laurent expansion of f in the neighborhood of a singularity z_0 has only a finite number of negative terms,

$$f(z) = \frac{a_{-m}}{(z - z_0)^m} + \cdots + a_0 + a_1(z - z_0) + \cdots,$$

and $a_{-m} \neq 0$. Then f is said to have a **pole of order** (or multiplicity m) at z_0. However, we still say that the **order of f at z_0 is** $-m$, that is,

$$\operatorname{ord}_{z_0} f = -m,$$

because we want the formula

$$\operatorname{ord} z_0(fg) = \operatorname{ord}_{z_0} f + \operatorname{ord}_{z_0} g$$

to be true. This situation is characterized as follows:

f has a pole of order m at z_0 if and only if $f(z)(z - z_0)^m$ is holomorphic at z_0 and has no zero at z_0.

The proof is immediate and is left to the reader.

If g is holomorphic at z_0 and $g(z_0) \neq 0$, then the function f defined by

$$f(z) = (z - z_0)^{-m} g(z)$$

in a neighborhood of z_0 from which z_0 is deleted, has a pole of order m. We abide by the convention that a pole is a zero of negative order.

A pole of order 1 is said to be a **simple pole**.

Examples. The function $1/z$ has a simple pole at the origin.

The function $1/\sin z$ has a simple pole at the origin. This comes from the power series expansion, since

$$\sin z = z(1 + \text{higher terms}),$$

and

$$\frac{1}{\sin z} = \frac{1}{z}(1 + \text{higher terms})$$

by inverting the series $1/(1 - h) = 1 + h + h^2 + \cdots$ for $|h| < 1$.

Let f be defined on an open set U except at a discrete set of points S which are poles. Then we say that f is **meromorphic** on U. If z_0 is such a point, then there exists an integer m such that $(z - z_0)^m f(z)$ is holomorphic in a neighborhood of z_0. Thus f is the quotient of two holomorphic functions in the neighborhood of a point. We say that f is **meromorphic at a point** z_0 if f is meromorphic on some open set U containing z_0.

Example. Let $P(z)$ be a polynomial. Then $f(z) = 1/P(z)$ is a meromorphic function. This is immediately seen by factoring $P(z)$ into linear factors.

Example. A meromorphic function can be defined often by a uniformly convergent series. For instance, let

$$f(z) = \frac{1}{z} + \sum_{n=1}^{\infty} \frac{z}{z^2 - n^2}.$$

We claim that f is meromorphic on **C** and has simple poles at the integers, but is holomorphic elsewhere.

We prove that f has these properties inside every disc of radius R centered at the origin. Let $R > 0$ and let $N > 2R$. Write

$$f(z) = g(z) + h(z),$$

where

$$g(z) = \frac{1}{z} + \sum_{n=1}^{N} \frac{z}{z^2 - n^2} \quad \text{and} \quad h(z) = \sum_{n=N+1}^{\infty} \frac{z}{z^2 - n^2}.$$

Then g is a rational function, and is therefore meromorphic on **C**. Furthermore, from its expression as a finite sum, we see that g has simple poles at the integers n such that $|n| \leq N$.

For the infinite series defining h, we apply Theorem 1.5 and prove that the series is uniformly convergent. For $|z| < R$ we have the estimate

$$\left| \frac{z}{z^2 - n^2} \right| \leq \frac{R}{n^2 - R^2} = \frac{1}{n^2} \frac{R}{1 - (R/n)^2}.$$

The denominator satisfies

$$1 - (R/n)^2 \geq \tfrac{3}{4}$$

for $n > N > 2R$. Hence

$$\left| \frac{z}{z^2 - n^2} \right| \leq \frac{4R}{3n^2} \qquad \text{for} \quad n \leq 2R.$$

Therefore the series for h converges uniformly in the disc $|z| < R$, and h is holomorphic in this disc. This proves the desired assertion.

Essential Singularities

If the Laurent series has a infinite number of negative terms, then we say that z_0 is an **essential singularity** of f.

Example. The function $f(z) = e^{1/z}$ has an essential singularity at $z = 0$ because its Laurent series is

$$\sum_{n=0}^{\infty} \frac{1}{z^n n!}$$

Theorem 3.2 (Casorati–Weierstrass). *Let 0 be an essential singularity of the function f, and let D be a disc centered at 0 on which f is holomorphic except at 0. Let U be the complement of 0 in D. Then $f(U)$ is dense in the complex numbers. In other words, the values of f on U come arbitrarily close to any complex number.*

Proof. Suppose the thorem is false. There exists a complex number α and a positive number $s > 0$ such that

$$|f(z) - \alpha| > s \qquad \text{for all} \quad z \in U.$$

The function

$$g(z) = \frac{1}{f(z) - \alpha}$$

is then holomorphic on U, and bounded on the disc D. Hence 0 is a removable singularity of g, and g may be extended to a holomorphic function on all of D. It then follows that $1/g(z)$ has at most a pole at 0, which means that $f(z) - \alpha$ has at most a pole, contradicting the hypothesis that $f(z)$ has an essential singularity (infinitely many terms of negative order in its Laurent series). This proves the theorem.

Actually, it was proved by Picard that f not only comes arbitrarily close to every complex number, but takes on every complex value except possibly one. The function $e^{1/z}$ omits the value 0, so it is necessary to allow for this one omission.

We recall that an **analytic isomorphism**

$$f: U \to V$$

from one open set to another is an analytic function such that there exists another analytic function

$$g: V \to U$$

satisfying

$$f \circ g = \mathrm{id}_V \quad \text{and} \quad g \circ f = \mathrm{id}_U,$$

where id is the identity function. An **analytic automorphism** of U is an analytic isomorphism of U with itself.

Using the Casorati–Weierstrass theorem, we shall prove:

Theorem 3.3. *The only analytic automorphisms of* $\mathbf{C}$ *are the functions of the form* $f(z) = az + b$, *where* a, b *are constants,* $a \neq 0$.

Proof. Let f be an analytic automorphism of $\mathbf{C}$, let

$$f(z) = \sum_{n=0}^{\infty} a_n z^n$$

be the power series expansion for f, and let $w = 1/z$, so that we can define a function

$$h(z) = f(1/z) = \sum_{n=0}^{\infty} a_n (1/z)^n$$

for $z \neq 0$. The function h has an isolated singularity at $z = 0$. If this singularity is essential, then the Casorati–Weierstrass theorem implies that its values for z near 0 come arbitrarily close to any point, and in particular, close to 0. However, let D be the unit disc and D^c the closed

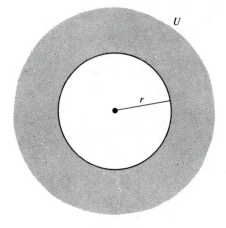

Figure 2

unit disc. Since the inverse function g of f is open, it follows that $g(D)$ is open, and $g(D^c)$ is compact, hence contained in a circle of radius R for sufficiently large R. Let U be the outside of the circle, as on Fig. 2. Then U is open, and $f(1/z)$ for $z \in U$ does not lie in D. This contradicts Casorati–Weierstrass, and proves that h cannot have an essential singularity at 0.

Therefore the series for h, hence for f, has only a finite number of terms, and

$$f(z) = a_0 + a_1 z + \cdots + a_N z^N$$

is a polynomial of degree N for some N. If f has two distinct roots, then f cannot be injective, contradicting the fact that f has an inverse function. Hence f has only one root, and

$$f(z) = a(z - z_0)^N$$

for some z_0. If $N > 1$, it is then clear that f is not injective so we must have $N = 1$, and the theorem is proved.

EXERCISES V §3

1. Show that the following series define a meromorphic function on $\mathbf{C}$ and determine the set of poles, and their orders.

(a) $\displaystyle\sum_{n=0}^{\infty} \frac{(-1)^n}{n!(n+z)}$ (b) $\displaystyle\sum_{n=1}^{\infty} \frac{1}{z^2 + n^2}$ (c) $\displaystyle\sum_{n=1}^{\infty} \frac{1}{(z+n)^2}$

(d) $\displaystyle\sum_{n=1}^{\infty} \frac{\sin nz}{n!(z^2 + n^2)}$ (e) $\displaystyle\frac{1}{z} + \sum_{\substack{n \neq 0 \\ n=-\infty}}^{\infty} \left[\frac{1}{z-n} + \frac{1}{n} \right]$

2. Show that the function

$$f(z) = \sum_{n=1}^{\infty} \frac{z^2}{n^2 z^2 + 8}$$

is defined and continuous for all real values of z. Determine the region of the complex plane in which this function is analytic. Determine its poles.

3. Let f be a non-constant entire function, i.e. a function analytic on all of $\mathbf{C}$. Show that the image of f is dense in $\mathbf{C}$.

4. Let f be meromorphic on an open set U. Let

$$\varphi: V \to U$$

be an analytic isomorphism. Suppose $\varphi(z_0) = w_0$, and f has order n at w_0. Show that $f \circ \varphi$ has order n at z_0. In other words, the order is invariant under analytic isomorphisms. [Here n is a positive or negative integer.]

5. **The Riemann Sphere.** Let S be the union of $\mathbf{C}$ and a single other point denoted by ∞, and called infinity. Let f be a function on S. Let $t = 1/z$, and define

$$g(t) = f(1/t)$$

for $t \neq 0, \infty$. We say that f has an **isolated singularity** resp. is **meromorphic** resp. is **holomorphic) at infinity** if g has an isolated singularity (resp. is meromorphic, resp. is holomorphic) at 0. The order of g at 0 will also be called the **order of f at infinity**. If g has a removable singularity at 0, and so can be defined as a holomorphic function in a neighborhood of 0, then we say that f is **holomorphic at infinity**.

We say that f is meromorphic on S if f is meromorphic on $\mathbf{C}$ and is also meromorphic at infinity. We say that f is holomorphic on S if f is holomorphic on $\mathbf{C}$ and is also holomorphic at infinity.
Prove:

The only meromorphic functions on S are the rational functions, that is, quotients of polynomials. The only holomorphic functions on S are the constants. If f is holomorphic on $\mathbf{C}$ and has a pole at infinity, then f is a polynomial.

In this last case, how would you describe the order of f at infinity in terms of the polynomial?

6. Let f be a meromorphic function on the Riemann sphere, so a rational function by Exercise 2. Prove that

$$\sum_{P} \operatorname{ord}_P f = 0,$$

where the sum is taken over all points P which are either points of $\mathbf{C}$, or $P = \infty$.

7. Let P_i $(i = 1, \ldots, r)$ be points of $\mathbf{C}$ or ∞, and let m_i be integers such that

$$\sum_{i=1}^{r} m_i = 0.$$

Prove that there exists a meromorphic function f on the Riemann sphere such that

$$\operatorname{ord}_{P_i} f = m_i \quad \text{for} \quad i = 1, \ldots, r$$

and $\operatorname{ord}_P f = 0$ if $P \neq P_i$.

V §4. Dixon's Proof of Cauchy's Theorem

It is possible to give an alternative proof of the global version of Cauchy's theorem which bypasses the topological considerations of Chapter IV, §3, and replaces them by analysis. After the homotopy form of Cauchy's theorem is obtained, one gets immediately Cauchy's formula for a circle. One can then obtain Liouville's theorem that an entire bounded function is constant. Using only these analytic facts, we now give Dixon's proof. An instructor or reader can easily reorder the material by striving for these analytic facts immediately after Chapter III if his taste so inclines him.

We define a function g on $U \times U$ by:

$$g(z, w) = \begin{cases} \dfrac{f(w) - f(z)}{w - z} & \text{if} \quad w \neq z, \\ f'(z) & \text{if} \quad w = z. \end{cases}$$

Then g is continuous on $U \times U$. This is obvious for points off the diagonal, and if (z_0, z_0) is on the diagonal, then

$$g(z, w) - g(z_0, z_0) = \frac{1}{w - z} \int_z^w [f'(\zeta) - f'(z_0)] \, d\zeta$$

for (z, w) close to (z_0, z_0). The integral can be taken along the line segment from z to w. Estimating the right-hand side, we see that $1/|w - z|$ cancels the length of the interval, and the expression under the integral sign tends to 0 by the continuity of f', as (z, w) approaches (z_0, z_0). Thus g is continuous.

We now define a function h on $\mathbf{C}$ by two integrals:

$$h(z) = \frac{1}{2\pi i} \int_\gamma g(z, w) \, dw \quad \text{if} \quad z \in U,$$

$$h(z) = \frac{1}{2\pi i} \int_\gamma \frac{f(w)}{w - z} \, dw \quad \text{if} \quad z \notin U.$$

We shall prove that h is a bounded entire function, whence constant by Liouville's theorem, whence equal to 0 by letting z tend to infinity for $z \notin U$, and using the definition of h. It is then clear that for $z \in U$ the first integral being zero immediately implies Cauchy's formula

$$\frac{1}{2\pi i} \int_\gamma \frac{f(w)}{w - z} \, dw = f(z) W(\gamma, z).$$

We have already seen in Remark 1 following Theorem 2.5 in Chapter IV, that Cauchy's formula implies Cauchy's theorem.

There remains therefore to prove that h is an analytic function and is bounded. As for the boundedness, suppose that z lies outside a large circle. Then

$$\int_\gamma \frac{f(z)}{w - z} \, dw = f(z) \int_\gamma \frac{1}{w - z} \, dw = 0$$

because the winding number of γ with respect to z is 0 by Lemma 1.3 of Chapter IV. Furthermore, if $z \notin U$, then

$$\int_\gamma \frac{f(w)}{w - z} \, dw$$

is bounded since $f(w)$ is bounded for w on γ, and the distance between γ and z is bounded from below, by the distance between γ (which is compact) and the complement of U. Finally, if $z \in U$ and z is inside a large circle, then the function $g(z, w)$ is bounded (being continuous). This proves that h is bounded.

The only problem about the analyticity of h is at a point of γ or at a boundary point of U. Let z_0 be a boundary point. Then z_0 does not lie on γ, and since γ is closed, there is a disc centered at z_0 which does not intersect γ. Such a disc intersects the complement of U, i.e. contains points not in U. By Lemma 1.2 of Chapter IV we conclude that $W(\gamma, z_0) = 0$, and therefore that the two integrals expressing h according as $z \in U$ or $z \notin U$ actually are equal to the same expression

$$\frac{1}{2\pi i} \int \frac{f(w)}{w - z} \, dw$$

for z in the disc. This proves that h is analytic at every point of the boundary.

From the uniform continuity of g on compact subsets of $U \times U$ it follows at once that h is continuous. To prove that h is analytic, by Theorem 3.2 of Chapter III, it suffices to prove that in some disc centered at z_0, the integral of h around the boundary of any rectangle contained in the disc is 0. But we have

$$\int_{\partial R} h(z) \, dz = \frac{1}{2\pi i} \int_{\partial R} \int_{\gamma} g(z, w) \, dw \, dz$$

$$= \frac{1}{2\pi i} \int_{\gamma} \int_{\partial R} g(z, w) \, dz \, dw.$$

Since for each w, the function $z \mapsto g(z, w)$ is analytic, we obtain the value 0, thereby concluding the proof.

CHAPTER VI

Calculus of Residues

We have established all the theorems needed to compute integrals of analytic functions in terms of their power series expansions. We first give the general statements covering this situation, and then apply them to examples.

VI §1. The Residue Formula

Let

$$f(z) = \sum_{n=-\infty}^{\infty} a_n(z - z_0)^n$$

have a Laurent expansion at a point z_0. We call a_{-1} the **residue** of f at z_0, and write

$$a_{-1} = \text{Res}_{z_0} f.$$

Theorem 1.1. *Let z_0 be an isolated singularity of f, and let C be a small circle centered at z_0 such that f is holomorphic on C and its interior, except possibly at z_0. Then*

$$\int_C f(\zeta)\, d\zeta = 2\pi i a_{-1} = 2\pi i\, \text{Res}_{z_0} f.$$

Proof. Since the series for $f(\zeta)$ converges uniformly and absolutely for ζ on the circle, we may integrate it term by term. The integral of $(\zeta - z_0)^n$ over the circle is equal to 0 for all values of n except possibly when $n = -1$, in which case we know that the value is $2\pi i$, cf. Examples 1 and 4 of Chapter III, §2. This proves the theorem.

From this local result, we may then deduce a global result for more general paths, by using the reduction of Theorem 2.2, Chapter IV.

Theorem 1.2 (Residue Formula). *Let U be an open set, and γ a closed chain in U such that γ is homologous to 0 in U. Let f be analytic on U except at a finite number of points $z_1, \ldots, z_n$. Let $m_i = W(\gamma, z_i)$. Then*

$$\int_\gamma f = 2\pi\sqrt{-1} \sum_{i=1}^{n} m_i \cdot \operatorname{Res}_{z_i} f.$$

Proof. Immediate by plugging Theorem 1.1 in the above mentioned theorem of Chapter IV.

Theorem 1.2 is used most often when U is simply connected, in which case every closed path is homologous to 0 in U, and the hypothesis on γ need not be mentioned explicitly. In the applications, U will be a disc, or the inside of a rectangle, where the simple connectedness is obvious.

Remark. The notation $\sqrt{-1}$ is the standard device used when we don't want to confuse the complex number i with an index i.

We shall give examples how to find residues.

A pole of a function f is said to be **simple** if it is of order 1, in which case the power series expansion of f is of type

$$f(z) = \frac{a_{-1}}{z - z_0} + a_0 + \text{higher terms},$$

and $a_{-1} \neq 0$.

Lemma 1.3. *Let f have a simple pole at z_0, and let g be holomorphic at z_0. Then*

$$\operatorname{Res}_{z_0}(fg) = g(z_0)a_{-1}.$$

Proof. Say $z_0 = 0$. We have

$$f(z)g(z) = \left(\frac{a_{-1}}{z} + \cdots\right)(b_0 + b_1 z + \cdots)$$

$$= \frac{a_{-1}b_0}{z} + \text{higher terms},$$

so our assertion is clear.

Example. Find the residue of $f(z) = \dfrac{z^2}{z^2 - 1}$ at $z = 1$.

To do this, we write

$$f(z) = \frac{z^2}{(z + 1)(z - 1)}.$$

Note that $g(z) = z^2/(z + 1)$ is holomorphic at 1, and that the residue of $1/(z - 1)$ is 1. Hence

$$\text{Res}_1 f = g(1) = 1/2.$$

Example. Find the residue of $(\sin z)/z^2$ at $z = 0$. We have

$$\frac{\sin z}{z^2} = \frac{1}{z^2}\left(z - \frac{z^3}{3!} + \cdots\right)$$

$$= \frac{1}{z} + \text{higher terms.}$$

Hence the desired residue is 1.

Example. Find the residue of $f(z) = \dfrac{z^2}{(z + 1)(z - 1)^2}$ at $z = 1$.

We note that the function

$$g(z) = \frac{z^2}{z + 1}$$

is holomorphic at $z = 1$, and has an expansion of type

$$g(z) = b_0 + b_1(z - 1) + \text{higher terms.}$$

Then

$$f(z) = \frac{g(z)}{(z - 1)^2} = \frac{b_0}{(z - 1)^2} + \frac{b_1}{(z - 1)} + \cdots$$

and therefore the residue of f at 1 is b_1, which we must now find. We write $z = 1 + (z - 1)$, so that

$$\frac{z^2}{z + 1} = \frac{1 + 2(z - 1) + (z + 1)^2}{2(1 + \frac{1}{2}(z - 1))}.$$

Inverting by the geometric series gives

$$\frac{z^2}{z+1} = \frac{1}{2}\left(1 + \frac{3}{2}(z-1) + \cdots\right).$$

Therefore

$$f(z) = \frac{1}{2(z-1)^2} + \frac{3/4}{z-1} + \cdots$$

whence $\text{Res}_1 f = 3/4$.

Example. Let C be a circle centered at 1, of radius 1. Let

$$f(z) = \frac{z^2}{(z+1)(z-1)^2}.$$

Find $\int_C f$.

The function f has only two singularities, at 1 and -1, and the circle is contained in a disc of radius > 1, centered at 1, on which f is holomorphic except at $z = 1$. Hence the residue formula and the preceding example give us

$$\int_C f = 2\pi i \cdot \tfrac{3}{4}.$$

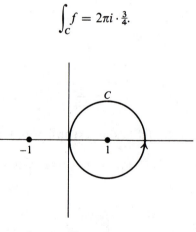

Figure 1

If C is the boundary of the rectangle as shown on Fig. 2, then we also find

$$\int_C f = 2\pi i \cdot \tfrac{3}{4}.$$

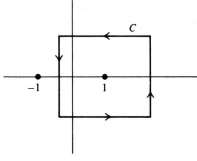

Figure 2

Example. Let $f(z) = z^2 - 2z + 3$. Let C be a rectangle as shown on Fig. 3, oriented clockwise. Find

$$\int_C \frac{1}{f(z)} \, dz.$$

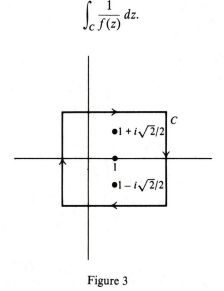

Figure 3

The roots of $f(z)$ are found by the quadratic formula to be

$$\frac{2 \pm \sqrt{-8}}{2}$$

and so are $z_1 = 1 + i\sqrt{2}$ and $z_2 = 1 - i\sqrt{2}$. The rectangle goes around these two points, in the clockwise direction. The residue of $1/f(z)$ at z_1 is $1/(z_1 - z_2)$ because f has a simple pole at z_1. The residue of $1/f(z)$ at z_2 is $1/(z_2 - z_1)$ for the same reason. The desired integral is equal to

$$-2\pi i(\text{sum of the residues}) = 0.$$

Example. Let f be the same function as in the preceding example, but now find the integral of $1/f$ over the rectangle as shown on Fig. 4. The rectangle is oriented clockwise.

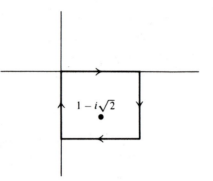

Figure 4

In this case, we have seen that the residue at $1 - i\sqrt{2}$ is

$$\frac{1}{z_2 - z_1} = \frac{1}{-2i\sqrt{2}}$$

Therefore the integral over the rectangle is equal to

$$-2\pi i(\text{residue}) = -2\pi i/(-2i\sqrt{2}) = \pi/\sqrt{2}.$$

Next we give an example which has theoretical significance, besides computational significance.

Example. Let f have a power series expansion with only a finite number of negative terms (so at most a pole), say at the origin,

$$f(z) = a_m z^m + \text{higher terms}, \qquad a_m \neq 0,$$

and m may be positive or negative. Then we can write

$$f(z) = a_m z^m(1 + h(z)),$$

where $h(z)$ is a power series with zero constant term. For any two functions f, g we know the derivative of the product,

$$(fg)' = f'g + fg',$$

so that dividing by fg yields

$$\boxed{\frac{(fg)'}{fg} = \frac{f'}{f} + \frac{g'}{g}.}$$

Therefore we find for $f(z) = (a_m z^m)(1 + h(z))$,

$$\frac{f'(z)}{f(z)} = \frac{m}{z} + \frac{h'(z)}{1 + h(z)}$$

and $h'(z)/(1 + h(z))$ is holomorphic at 0. Consequently, we get:

Lemma 1.4. *Let f be meromorphic at 0. Then*

$$\operatorname{Res}_0 f'/f = \operatorname{ord}_0 f$$

and for any point z_0 where f has at most a pole,

$$\operatorname{Res}_{z_0} f'/f = \operatorname{ord}_{z_0} f.$$

Theorem 1.5. *Let γ be a closed chain in U, homologous to 0 in U. Let f be meromorphic on U, with only a finite number of zeros and poles, say at the points $z_1, \ldots, z_n$, none of which lie on γ. Let $m_i = W(\gamma, z_i)$. Then*

$$\int_\gamma f'/f = 2\pi\sqrt{-1} \sum m_i \operatorname{ord}_{z_i} f.$$

Proof. This is immediate by plugging the statement of the lemma into the residue formula.

In applications, γ is frequently equal to a circle C, or a rectangle, and the points $z_1, \ldots, z_n$ are inside C. Suppose that the zeros of f inside C are

$$a_1, \ldots, a_r$$

and the poles are

$$b_1, \ldots, b_s.$$

Then in the case,

$$\boxed{\int_C f'/f = 2\pi\sqrt{-1} \left(\sum_{i=1}^r \operatorname{ord}_{a_i} f - \sum_{j=1}^s \operatorname{mul}_{b_j} f \right).}$$

We follow our convention whereby the multiplicity of a pole is the negative of the order of f at the pole, so that

$$\text{mul}_{b_j} f = -\text{ord}_{b_j} f$$

by definition.

If one counts zeros and poles with their multiplicities, one may re-
phrase the above formula in the more suggestive fashion:

*Let C be a simple closed curve, and let f be analytic on C and its interior.
Assume that f has no zero or pole on C. Then*

$$\int_C f'/f = 2\pi i \ (\textit{number of zeros} - \textit{number of poles}),$$

*where number of zeros = number of zeros of f in the interior of C, and
number of poles = number of poles of f in the interior of C.*

Of course, we have not proved that a simple closed curve has an
"interior". The theorem is applied in practice only when the curve is so
explicitly given (as with a circle or rectangle) that it is clear what
"interior" is meant.

Besides, one can (not so artificially) formalize what is needed of the
notion of "interior" so that one can use the standard language. Let γ be
a closed path. We say that γ has an **interior** if $W(\gamma, \alpha) = 0$ or 1 for every
complex number α which does not lie on γ. Then the set of points α such
that $W(\gamma, \alpha) = 1$ will be called the **interior** of γ. It's that simple.

Theorem 1.6 (Rouché's Theorem). *Let γ be a closed path homologous to
0 in U and assume that γ has an interior. Let f, g be analytic on U, and*

$$|f(z) - g(z)| < |f(z)|$$

*for z on γ. Then f and g have the same number of zeros in the interior
of γ.*

Proof. Note that the assumption implies automatically that f, g have
no zero on γ. We have

$$\left| \frac{g(z)}{f(z)} - 1 \right| < 1$$

for z on γ. Then the values of the function g/f are contained in the open
disc with center 1 and radius 1. Let $F = g/f$. Then $F \circ \gamma$ is a closed path
contained in that disc, and therefore

$$W(F \circ \gamma, 0) = 0$$

because 0 lies outside the disc. If γ is defined on $[a, b]$ then

$$0 = W(F \circ \gamma, 0) = \int_{F \circ \gamma} \frac{1}{z} dz = \int_a^b \frac{F'(\gamma(t))}{F(\gamma(t))} \gamma'(t) \, dt$$

$$= \int_\gamma F'/F$$

$$= \int_\gamma g'/g - f'/f.$$

What we want now follows from Theorem 1.5, as desired.

Example. Let $P(z) = z^8 - 5z^3 + z - 2$. We want to find the number of roots of this polynomial inside the unit circle. Let

$$f(z) = -5z^3.$$

For $|z| = 1$ it is immediate that

$$|f(z) - P(z)| = |-z^8 - z + 2| < |f(z)| = 5.$$

Hence f and P have the same number of zeros inside the unit circle, and this number is clearly equal to 3. (Remember, you have to count multiplicities, and the equation

$$5z^3 = 0$$

has one zero with multiplicity 3.)

We shall use Rouché's theorem to give an alternative treatment of the inverse function theorem, not depending on solving for an inverse power series as was done in Chapter II, §5.

Theorem 1.7. Let f be analytic in a neighborhood of a point z_0, and assume $f'(z_0) \neq 0$. Then f is a local analytic isomorphism at z_0.

Proof. Making translations, we may assume without loss of generality that $z_0 = 0$ and $f(z_0) = 0$, so that

$$f(z) = \sum_{n=m}^\infty a_n z^n,$$

and $m \geq 1$. Since $f'(0) = a_1$ we have $m = 1$ and $a_1 \neq 0$. Dividing by a_1 we may assume without loss of generality that $a_1 = 1$. Thus f has the power series expansion

$$f(z) = z + h(z),$$

where $h(z)$ is divisible by z^2. In particular, if we restrict the values of z to some sufficiently small disc around z_0, then there is a constant $K > 0$ such that

$$|f(z) - z| \leqq K|z|^2.$$

Let C_r be the circle of radius r, and let $|\alpha| < r/2$. Let r be sufficiently small, and let

$$f_\alpha(z) = f(z) - \alpha \qquad \text{and} \qquad g_\alpha(z) = z - \alpha.$$

We have the inequality

$$|f_\alpha(z) - g_\alpha(z)| = |f(z) - z| \leqq K|z|^2.$$

If z is on C_r, that is $|z| = r$, then

$$K|z|^2 = Kr^2 < |z - \alpha| = |g_\alpha(z)|$$

because $|z - \alpha| > r/2$ and $Kr^2 < r/2$ (for instance, taking $r < 1/2K$). By Rouché's theorem, f_α and g_α have the same number of zeros inside C_r, and since g_α has exactly one zero, it follows that f_α has exactly one zero. This means that the equation

$$f(z) = \alpha$$

has exactly one solution inside C_r if $|\alpha| < r/2$.

Let U be the set of points z inside C_r such that

$$|f(z)| < r/2.$$

Then U is open because f is continuous, and we have just shown that

$$f : U \to D(0, r/2)$$

is a bijection of U with the disc of radius $r/2$. The argument we have given also shows that f is an open mapping, and hence the inverse function

$$\varphi : D(0, r/2) \to U$$

is continuous. There remains only to prove that φ is analytic. As in freshman calculus, we write the Newton quotient

$$\frac{\varphi(w) - \varphi(w_1)}{w - w_1} = \frac{z - z_1}{f(z) - f(z_1)}.$$

Fix w_1 with $|w_1| < r/2$, and let w approach w_1. Since φ is continuous, it must be that $z = \varphi(w)$ approaches $z_1 = \varphi(w_1)$. Thus the right-hand side approaches

$$1/f'(z_1),$$

provided we took r so small that $f'(z_1) \neq 0$ for all z_1 inside the circle of radius r, which is possible by the continuity of f' and the fact that $f'(0) \neq 0$. This proves that φ is analytic and concludes the proof of the theorem.

EXERCISES VI §1

Find the residues of the following functions at 0.

1. $(z^2 + 1)/z$

2. $(z^2 + 3z - 5)/z^3$

3. $z^3/(z - 1)(z^4 + 2)$

4. $(2z + 1)/z(z^3 - 5)$

5. $(\sin z)/z^4$

6. $(\sin z)/z^5$

7. $(\sin z)/z^6$

8. $(\sin z)/z^7$

9. e^z/z

10. e^z/z^2

11. e^z/z^3

12. e^z/z^4

13. $z^{-2}\log(1 + z)$

14. $e^z/\sin z$

Find the residues of the following functions at 1.

15. $1/(z^2 - 1)(z + 2)$

16. $(z^3 - 1)(z + 2)/(z^4 - 1)^2$

17. Factor the polynomial $z^n - 1$ into factors of degree 1. Find the residue at 1 of $1/(z^n - 1)$.

18. Let $z_1, \dots, z_n$ be distinct complex numbers. Let C be a circle around z_1 such that C and its interior do not contain z_j for $j > 1$. Let

$$f(z) = (z - z_1)(z - z_2)\cdots(z - z_n).$$

Find

$$\int_C \frac{1}{f(z)}\, dz.$$

19. Find the residue at i of $1/(z^4 - 1)$. Find the integral

$$\int_C \frac{1}{(z^4 - 1)}\, dz$$

where C is a circle of radius $1/2$ centered at i.

20. (a) Find the integral

$$\int_C \frac{1}{z^2 - 3z + 5} \, dz,$$

where C is a rectangle oriented clockwise, as shown on the figure.

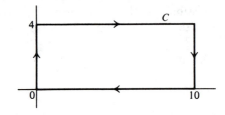

Figure 5

(b) Find the integral $\int_C 1/(z^2 + z + 1) \, dz$ over the same C.

(c) Find the integral $\int_C 1/(z^2 - z + 1) \, dz$ over this same C.

21. (a) Let f be holomorphic at z_0, and $f(z_0) = 0$. Prove that if $f'(z_0) \neq 0$, then f has a simple zero, i.e. $\mathrm{ord}_{z_0} f = 1$. Then $1/f(z)$ has a simple pole at z_0. Prove that the residue of $1/f(z)$ at z_0 is $1/f'(z_0)$.

(b) Let $z_1, \ldots, z_n$ be distinct complex numbers. Determine explicitly the partial fraction decomposition (i.e. the numbers a_i):

$$\frac{1}{(z - z_1) \cdots (z - z_n)} = \frac{a_1}{z - z_1} + \cdots + \frac{a_n}{z - z_n}.$$

(c) Let $P(z)$ be a polynomial of degree $\leq n - 1$, and let $a_1, \ldots, a_n$ be distinct complex numbers. Assume that there is a partial fraction decomposition of the form

$$\frac{P(z)}{(z - a_1) \cdots (z - a_n)} = \frac{c_1}{z - a_1} + \cdots + \frac{c_n}{z - a_n}.$$

Prove that

$$c_1 = \frac{P(a_1)}{(a_1 - a_2) \cdots (a_1 - a_n)},$$

and similarly for the other coefficients c_j.

22. Let g be a function holomorphic at a point z_0. Show that the residue of the function

$$\frac{g(z)}{z - z_0}$$

at z_0 is equal to $g(z_0)$. What is the relevance of this problem to the preceding exercise?

23. (a) Let f be a function which is analytic on the upper half plane, and on the real line. Assume that there exist numbers $B > 0$ and $c > 0$ such that

$$|f(\zeta)| \leq \frac{B}{|\zeta|^c}$$

for all ζ. Prove that for any z in the upper half plane, we have the integral formula

$$f(z) = \frac{1}{2\pi i} \int_{-\infty}^{\infty} \frac{f(t)}{t - z}\, dt. \qquad\qquad f(t)$$

Hint: Consider the integral over the path shown on the figure, and take the limit as $R \to \infty$.

The path consists of the segment from $-R$ to R on the real axis, and the semicircle S_R as shown.

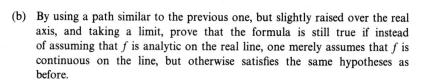

Figure 6

(b) By using a path similar to the previous one, but slightly raised over the real axis, and taking a limit, prove that the formula is still true if instead of assuming that f is analytic on the real line, one merely assumes that f is continuous on the line, but otherwise satisfies the same hypotheses as before.

24. Determine the poles and find the residues of the following functions. (a) $1/\sin z$ (b) $1/(1 - e^z)$ (c) $z/(1 - \cos z)$.

25. Show that

$$\int_{|z|=1} \frac{\cos e^{-z}}{z^2}\, dz = 2\pi i \cdot \sin 1.$$

26. Find the integrals, where C is the circle of radius 8 centered at the origin.

(a) $\displaystyle \int_C \frac{1}{\sin z} \, dz$

(b) $\displaystyle \int_C \frac{1}{1 - \cos z} \, dz$

(c) $\displaystyle \int_C \frac{1 + z}{1 - e^z} \, dz$

(d) $\displaystyle \int_C \tan z \, dz$

(e) $\displaystyle \int_C \frac{1 + z}{1 - \sin z} \, dz$

27. Let f be holomorphic on and inside the unit circle, $|z| \leq 1$, except for a pole of order 1 at a point z_0 on the circle. Let $f = \sum a_n z^n$ be the power series for f on the open disc. Prove that

$$\lim_{n \to \infty} \frac{a_n}{a_{n+1}} = z_0.$$

28. Let a be real > 1. Prove that the equation

$$ze^{a-z} = 1$$

has a single solution with $|z| \leq 1$, which is real and positive.

29. Let U be a connected open set, and let D be an open disc whose closure is contained in U. Let f be analytic on U and not constant. Assume that the absolute value $|f|$ is constant on the boundary of D. Prove that f has at least one zero in D.

30. Let f be a function analytic inside and on the unit circle. Suppose that $|f(z) - z| < |z|$ on the unit circle.
    (a) Show that $|f'(1/2)| \leq 8$.
    (b) Show that f has precisely one zero inside of the unit circle.

31. Determine the number of zeros of the polynomial

$$z^{87} + 36z^{57} + 71z^4 + z^3 - z + 1$$

inside the circle
    (a) of radius 1,
    (b) of radius 2, centered at the origin.
    (c) Determine the number of zeros of the polynomial

$$2z^5 - 6z^2 + z + 1 = 0$$

in the annulus $1 \leq |z| \leq 2$.

32. Let f, h be analytic on the closed disc of radius R, and assume that $f(z) \neq 0$ for z on the circle of radius R. Prove that there exists $\epsilon > 0$ such that $f(z)$ and $f(z) + \epsilon h(z)$ have the same number of zeros inside the circle of radius R.

Loosely speaking, we may say that f and a small perturbation of f have the same number of zeros inside the circle.

33. Let $f(z) = a_n z^n + \cdots + a_0$ be a polynomial with $a_n \neq 0$. Use Rouché's theorem to show that $f(z)$ and $a_n z^n$ have the same number of zeros in a disc of radius R for R sufficiently large.

34. Let f be analytic on $\mathbf{C}$ with the exception of a finite number of isolated singularities which may be poles. Define the **residue at infinity**

$$\operatorname{res}_\infty f = -\frac{1}{2\pi i} \int_{|z|=R} f(z)\, dz$$

for R so large that f has no singularities in $|z| \geq R$.
(a) Show that $\operatorname{res}_\infty f$ is independent of R.
(b) Show that the sum of the residues of f at all singularities and the residue at infinity is equal to 0.

35. **Cauchy's Residue Formula on the Riemann Sphere.** Recall Exercise 2 of Chapter V, §3 on the Riemann sphere. By a (meromorphic) **differential form** ω on the Riemann sphere S, we mean an expression of the form

$$\omega = f(z)\, dz,$$

where f is a rational function. For any point $z_0 \in \mathbf{C}$ the **residue** of ω at z_0 is defined to be the residue of f at z_0. For the point ∞, we write $t = 1/z$,

$$dt = -\frac{1}{z^2}\, dz \qquad \text{and} \qquad dz = -\frac{1}{t^2}\, dt,$$

so we write

$$\omega = f(1/t)\left(-\frac{1}{t^2}\right) dt = -\frac{1}{t^2} f(1/t)\, dt.$$

The **residue of** ω **at infinity** is then defined to be the residue of $-\dfrac{1}{t^2} f(1/t)$ at

$t = 0$. Prove:
(a) $\sum$ residues $\omega = 0$, if the sum is taken over all points of $\mathbf{C}$ and also infinity.
(b) Let γ be a circle of radius R centered at the origin in $\mathbf{C}$. If R is sufficiently large, show that

$$\frac{1}{2\pi i} \int_\gamma f(z)\, dz = -\text{residue of } f(z)\, dz \text{ at infinity.}$$

(Instead of a circle, you can also take a simple closed curve such that all the poles of f in $\mathbf{C}$ lie in its interior.)
(c) If R is arbitrary, and f has no pole on the circle, show that

$$\frac{1}{2\pi i} \int_\gamma f(z)\, dz = -\sum \text{ residues of } f(z)\, dz \text{ outside the circle,}$$
$$\text{including the residue at } \infty.$$

[**Note**: In dealing with (a) and (b), you can either find a direct algebraic proof of (a), and deduce (b) from it, or you can prove (b) directly, using a change of variables $t = 1/z$, and then deduce (a) from (b). You probably should carry both ideas out completely to understand fully what's going on.]

VI §2. Evaluation of Definite Integrals

Let $f(x)$ be a continuous function of a real variable x. We want to compute

$$\int_{-\infty}^{\infty} f(x)\,dx = \lim_{R \to \infty} \int_{-R}^{R} f(x)\,dx.$$

We shall use the following method. We let γ be the closed path as indicated on Fig. 7, consisting of a segment on the real line, and a semi-circle.

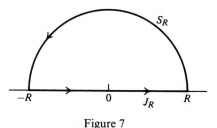

Figure 7

We suppose that $f(x)$ is the restriction to the line of a function f on the upper half plane, meromorphic and having only a finite number of poles. We let J_R be the segment from $-R$ to R, and let S_R be the semicircle. If we can prove that

$$\lim_{R \to \infty} \int_{S_R} f = 0$$

then by the residue formula, we obtain

$$\int_{-\infty}^{\infty} f(x)\,dx = 2\pi i \sum \text{residues of } f \text{ in the upper half plane.}$$

For this method to work, it suffices to know that $f(z)$ goes sufficiently fast to 0 when $|z|$ becomes large, so that the integral over the semicircle tends to 0 as the radius R becomes large. It is easy to state conditions under which this is true.

Theorem 2.1. *Suppose that there exists a number $B > 0$ such that for all $|z|$ sufficiently large, we have*

$$|f(z)| \leq B/|z|^2.$$

Then

$$\lim_{R \to \infty} \int_{S_R} f = 0$$

and the above formula is valid.

Proof. The integral is estimated by the sup norm of f, which is B/R^2 by assumption, multiplied by the length of the semicircle, which is πR. Since $\pi B/R$ tends to 0 as $R \to \infty$, our theorem is proved.

Remark. We really did not need an R^2, only R^{1+a} for some $a > 0$, so the theorem could be correspondingly strengthened.

Example. Let us compute

$$\int_{-\infty}^{\infty} \frac{1}{x^4 + 1} \, dx.$$

The function $1/(z^4 + 1)$ is meromorphic on $\mathbf{C}$, and its poles are at the zeros of $z^4 + 1$, that is the solutions of $z^4 = -1$, which are

$$e^{\pi i k/4}, \qquad k = 1, -1, 3, -3.$$

Let $f(z) = z^4 + 1$. Since $f'(z) = 4z^3 \neq 0$ unless $z = 0$, we conclude that all the zeros of f are simple. The two zeros in the upper half plane are

$$z_1 = e^{\pi i/4} \qquad \text{and} \qquad z_2 = e^{3\pi i/4}.$$

The residues of $1/f(z)$ at these points are $1/f'(z_1)$, $1/f'(z_2)$, respectively, and

$$f'(z_1) = 4z_1^3 = 4e^{3\pi i/4}, \qquad f'(z_2) = 4z_2^3 = 4e^{\pi i/4}.$$

The estimate

$$\left| \frac{1}{z^4 + 1} \right| \leq B/R^4$$

is satisfied for some constant B when $|z| = R$. Hence the theorem applies, and

$$\int_{-\infty}^{\infty} \frac{1}{x^4 + 1}\, dx = 2\pi i(\tfrac{1}{4}e^{-3\pi i/4} + \tfrac{1}{4}e^{-\pi i/4})$$

$$= \frac{\pi i}{2}\, e^{-\pi i/4}(e^{-2\pi i/4} + 1)$$

$$= \frac{\pi i}{2}\left(\frac{1 - i}{\sqrt{2}}\right)(1 - i)$$

$$= \frac{\pi}{\sqrt{2}}.$$

The estimate for $1/(z^4 + 1)$ on a circle of radius R presented no subtlety. We give an example where the estimate takes into account a different phenomenon, and a different path. The fact that the integral over the part going to infinity like the semicircle tends to 0 will be due to a more conditional convergence, and the evaluation of an integral explicitly.

Fourier Transforms

Integrals of the form discussed in the next examples are called **Fourier transforms,** and the technique shows how to evaluate them.

Theorem 2.2. *Let f be meromorphic on $\mathbf{C}$, having only a finite number of poles, not lying on the real axis. Suppose that there is a constant K such that*

$$|f(z)| \le K/|z|$$

for all sufficiently large $|z|$. Let $a > 0$. Then

$$\int_{-\infty}^{\infty} f(x)e^{iax}\, dx = 2\pi i \sum \text{residues of } e^{iaz}f(z) \text{ in the upper half plane.}$$

Proof. For simplicity, take $a = 1$. We integrate over any rectangle as shown on Fig. 8, taking $T = A + B$. Taking $A, B > 0$ sufficiently large, it suffices to prove that the integral over the three sides other than the bottom side tend to 0 as A, B tend to infinity. Note that

$$e^{iz} = e^{i(x + iy)} = e^{ix}e^{-y}.$$

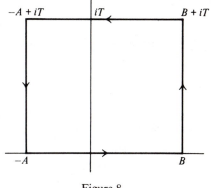

Figure 8

In absolute value this is e^{-y}, and tends to 0 rapidly as y tends to infinity. We show that the integral over the top tends to 0. Parametrizing the top by $x + iT$, with $-A \leq x \leq B$, we find

$$- \int_{\text{top}} e^{iz} f(z) \, dz = \int_{-A}^{B} e^{ix} e^{-T} f(x + iT) \, dx$$

and in absolute value, this is less than

$$e^{-T} \int_{-A}^{B} |f(x + iT)| \, dx \leq e^{-T} \frac{K}{T} (A + B).$$

Having picked $T = A + B$ shows that this integral becomes small as A, B become large, as desired.

For the right-hand side, we pick the parametrization

$$B + iy, \qquad \text{with} \quad 0 \leq y \leq T,$$

and we find that the right-hand side integral is bounded by

$$\left| \int_{0}^{T} e^{iB} e^{-y} f(B + iy) \, dy \right| \leq \frac{K}{B} \int_{0}^{T} e^{-y} \, dy = \frac{K}{B} (1 - e^{-T}),$$

which tends to 0 as B becomes large. A similar estimate shows that the integral over the left side tends to 0, and proves what we wanted.

Next we show an adjustment of the above techniques when the function may have some singularity on the real axis. We do this by an example.

Example. Let us compute

$$I = \int_0^\infty \frac{\sin x}{x}\, dx = \frac{1}{2}\int_{-\infty}^\infty \frac{\sin x}{x}\, dx.$$

$$= \frac{1}{2i}\lim_{\epsilon \to 0}\left[\int_{-\infty}^{-\epsilon}\frac{e^{ix}}{x}\, dx + \int_\epsilon^\infty \frac{e^{ix}}{x}\, dx\right].$$

We use the closed path C as shown on Fig. 9.

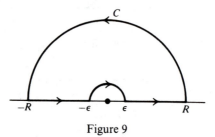

Figure 9

Let $S(\epsilon)$ be the small semicircle from ϵ to $-\epsilon$, oriented counterclock-wise, and let $S(R)$ be the big semicircle from R to $-R$ similarly oriented. The function e^{iz}/z has no pole inside C, and consequently

$$0 = \int_C e^{iz}/z\, dz = \int_{S(R)} + \int_{-R}^{-\epsilon} - \int_{S(\epsilon)} + \int_\epsilon^R e^{iz}/z\, dz.$$

Hence

$$\int_{-R}^{-\epsilon} + \int_\epsilon^R e^{iz}/z\, dz = \int_{S(\epsilon)} e^{iz}/z\, dz - \int_{S(R)} e^{iz}/z\, dz$$

$$= I_{S(\epsilon)} - I_{S(R)}.$$

We now assert that

$$\lim_{R \to \infty} I_{S(R)} = 0.$$

Proof. We have for $z = R(\cos\theta + i\sin\theta)$,

$$I_{S(R)} = \int_0^\pi \frac{e^{iR\cos\theta}e^{-R\sin\theta}}{iRe^{i\theta}}\, Rie^{i\theta}\, d\theta$$

so that

$$|I_{S(R)}| \leqq \int_0^\pi e^{-R\sin\theta}\, d\theta = 2\int_0^{\pi/2} e^{-R\sin\theta}\, d\theta.$$

But if $0 \leq \theta \leq \pi/2$, then $\sin \theta \geq 2\theta/\pi$, (any similar estimate would do), and hence

$$|I_{S(R)}| \leq 2 \int_0^{\pi/2} e^{-2R\theta/\pi} \, d\theta$$

$$= \frac{\pi}{R} (1 - e^{-R})$$

by Freshman calculus. This proves our assertion.

There remains to evaluate the limit of $I_{S(\epsilon)}$ as $\epsilon \to 0$. We state this as a general lemma.

Lemma. *Let g have a simple pole at 0. Then*

$$\lim_{\epsilon \to 0} \int_{S(\epsilon)} g(z) \, dz = \pi i \, \mathrm{Res}_0(g).$$

Proof. Write

$$g(z) = \frac{a}{z} + h(z),$$

where h is holomorphic at 0. Then the integral of h over $S(\epsilon)$ approaches 0 as $\epsilon \to 0$ because the length of $S(\epsilon)$ approaches 0 and h is bounded near the origin. A direct integration of a/z shows that the integral of a/z over the semicircle is equal to $\pi i a$. This proves the lemma.

We may therefore put everything together to find the value

$$\boxed{\int_0^\infty \frac{\sin x}{x} \, dx = \pi/2.}$$

Trigonometric Integrals

We wish to evaluate an integral of the form

$$\int_0^{2\pi} Q(\cos \theta, \sin \theta) \, d\theta,$$

where Q is a rational function of two variables, $Q = Q(x, y)$, which we assume is continuous on the unit circle. Since

$$\cos \theta = \frac{e^{i\theta} + e^{-i\theta}}{2} \qquad \text{and} \qquad \sin \theta = \frac{e^{i\theta} - e^{-i\theta}}{2i},$$

we see that these expressions are equal to

$$\frac{z + 1/z}{2} \quad \text{and} \quad \frac{z - 1/z}{2i},$$

respectively, when z lies on the unit circle, $z = e^{i\theta}$. It is therefore natural to consider the function

$$f(z) = \frac{Q\left(\frac{1}{2}\left(z + \frac{1}{z}\right), \frac{1}{2i}\left(z - \frac{1}{z}\right)\right)}{iz}$$

(the denominator iz is put there for a purpose which will become apparent in a moment). This function f is a rational function of z, and in view of our assumption on Q, it has no pole on the unit circle.

Theorem 2.3. *Let $Q(x, y)$ be a rational function which is continuous when $x^2 + y^2 = 1$. Let $f(z)$ be as above. Then*

$$\int_0^{2\pi} Q(\cos\theta, \sin\theta) \, d\theta = 2\pi i \sum \text{residues of } f \text{ inside the unit circle.}$$

Proof. Let C be the unit circle. Then

$$\int_C f(z) \, dz = 2\pi i \sum \text{residues of } f \text{ inside the circle.}$$

On the other hand, by definition the integral on the left is equal to

$$\int_0^{2\pi} f(e^{i\theta}) i e^{i\theta} \, d\theta = \int_0^{2\pi} Q(\cos\theta, \sin\theta) \, d\theta,$$

as desired. [The term iz in the denominator of f was introduced to cancel $ie^{i\theta}$ at this point.]

Example. Let us compute the integral

$$I = \int_0^{2\pi} \frac{1}{a + \sin\theta} \, d\theta$$

where a is real > 1. By the theorem,

$$I = 2\pi \sum \text{residues of } \frac{2i}{z^2 + 2iaz - 1} \text{ inside circle.}$$

The only pole inside the circle is at

$$z_0 = -ia + i\sqrt{a^2 - 1}$$

and the residue is

$$\frac{i}{z_0 + ia} = \frac{1}{\sqrt{a^2 - 1}}.$$

Consequently,

$$I = \frac{2\pi}{\sqrt{a^2 - 1}}.$$

Mellin Transforms

We give a final example introducing new complications. Integrals of type

$$\int_0^\infty f(x)x^a \frac{dx}{x}$$

are called **Mellin transforms** (they can be viewed as functions of a). We wish to show how to evaluate them. *We assume that $f(z)$ is analytic on* **C** *except for a finite number of poles, none of which lies on the positive real axis $0 < x$, and we also assume that a is not an integer.* Then under appropriate conditions on the behavior of f near 0, and when x becomes large, we can show that the following formula holds:

$$\int_0^\infty f(x)x^a \frac{dx}{x} = -\frac{\pi e^{-\pi i a}}{\sin \pi a} \sum \text{residues of } f(z)z^{a-1} \text{ at the poles of } f, \text{ excluding the residue at } 0.$$

We comment right away on what we mean by z^{a-1}, namely z^{a-1} is defined as

$$z^{a-1} = e^{(a-1)\log z},$$

where the log is defined on the simply connected set equal to the plane from which the axis $x \geq 0$ has been deleted. We take the value for the log such that if $z = re^{i\theta}$ and $0 < \theta < 2\pi$, then

$$\log z = \log r + i\theta.$$

Then, for instance,

$$\log i = \pi i/2 \quad \text{and} \quad \log(-i) = 3\pi i/2.$$

Precise sufficient conditions under which the formula is true are given in the next theorem. They involve suitable estimates for the function f near 0 and infinity.

Theorem 2.4. *The formula given for the integral*

$$\int_0^\infty f(x)x^a \frac{dx}{x}, \qquad \text{with} \quad a > 0,$$

is valid under the following conditions:

1. *There exists a number $b > a$ such that*

$$|f(z)| \ll 1/|z|^b \qquad for \quad |z| \to \infty.$$

2. *There exists a number b' with $0 < b' < a$ such that*

$$|f(z)| \ll 1/|z|^{b'} \qquad for \quad |z| \to 0.$$

The symbol $\ll$ means that the left-hand side is less than or equal to some constant times the right-hand side.

For definiteness, we carry out the arguments on a concrete example, and let the reader verify that the arguments work under the conditions stated in Theorem 2.4.

Example. We shall evaluate for $0 < a < 2$:

$$\int_0^\infty \frac{1}{x^2 + 1} x^a \frac{dx}{x} = \frac{\pi \cos a\pi/2}{\sin a\pi}.$$

We choose the closed path C as on Fig. 10. Then C consists of two line segments L^+ and L^-, and two pieces of semicircles $S(R)$ and $-S(\epsilon)$, if we take $S(\epsilon)$ oriented in counterclockwise direction. The angle φ which the two segments L^+ and L^- make with the positive real axis will tend to 0.

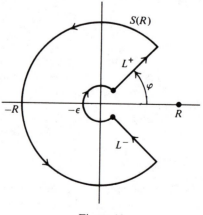

Figure 10

We let

$$g(z) = \frac{1}{z^2 + 1} z^{a-1}.$$

Then $g(z)$ has only simple poles at $z = i$ and $z = -i$, where the residues are found to be:

$$\text{at } i: \frac{1}{2i} e^{(a-1)\log i} = \frac{1}{2i} e^{(a-1)\pi i/2}$$

$$\text{at } -i: -\frac{1}{2i} e^{(a-1)\log(-i)} = -\frac{1}{2i} e^{(a-1)3\pi i/2}.$$

The sum of the residues inside C is therefore equal to

$$\frac{1}{2i} (e^{(a-1)\pi i/2} - e^{(a-1)3\pi i/2}) = -e^{a\pi i} \cos(a\pi/2),$$

after observing that $e^{\pi i/2} = i$, $e^{-3\pi i/2} = i$, and factoring out $e^{a\pi i}$ from the sum.

The residue formula yields

$$2\pi i \sum \text{residues} = I_{S(R)} + I_{L^-} - I_{S(\epsilon)} + I_{L^+},$$

where I_X denotes the integral of $f(z)$ over the path X. We shall prove:

The integrals $I_{S(R)}$ and $I_{S(\epsilon)}$ tend to 0 as R becomes large and ϵ becomes small, independently of the angle φ.

Proof. First estimate the integral over $S(R)$. When comparing functions of R, it is useful to use the following notation. Let $F(R)$ and $G(R)$ be functions of R, and assume that $G(R)$ is >0 for all R sufficiently large. We write

$$F(R) \ll G(R) \qquad (\text{for } R \to \infty)$$

if there exists a constant K such that

$$|F(R)| \leq KG(R)$$

for all R sufficiently large.
With this notation, using $z^{a-1} = e^{(a-1)\log z}$, and

$$|\log z| \leq \log R + \theta \leq \log R + 2\pi,$$

we find

$$|z^{a-1}| = |e^{(a-1)\log z}| \ll R^{a-1}.$$

Consequently from $|1/(z^2 + 1)| \ll 1/R^2$ for $|z| = R$, we find

$$\left| \int_{S(R)} f(z)z^{a-1}\, dz \right| \ll 2\pi R \, \frac{1}{R^2} \, \max|z^{a-1}| \ll R^a/R^2.$$

Since we assumed that $a < 2$, the quotient R^a/R^2 approaches 0 as R becomes large, as desired. The estimate is independent of φ.

We use a similar estimating notation for functions of ϵ,

$$F(\epsilon) \ll G(\epsilon) \qquad (\text{for } \epsilon \to 0)$$

if there exists a constant K such that

$$|F(\epsilon) \leqq KG(\epsilon)$$

for all $\epsilon > 0$ sufficiently small. With this notation, for $|z| = \epsilon$, we have

$$|z^{a-1}| = |e^{(a-1)\log z}| \ll 1/\epsilon^{a-1}.$$

Hence

$$\left| \int_{S(\epsilon)} f(z)z^{a-1}\, dz \right| \ll 2\pi\epsilon/\epsilon^{a-1} \ll \epsilon^{2-a}.$$

Again since we assumed that $a < 2$, the right-hand side approaches 0 and ϵ tends to 0, as desired. The estimate is independent of φ.

There remains to analyze the sums of the integrals over L^+ and L^-. We parametrize L^+ by

$$z(r) = re^{i\varphi}, \qquad \epsilon \ll r \ll R,$$

so that $\log z(r) = \log r + i\varphi$. Then

$$\int_{L^+} f(z)e^{(a-1)\log z}\, dz = \int_\epsilon^R f(re^{i\varphi})e^{(a-1)(\log r + i\varphi)}e^{i\varphi}\, dr$$

$$= \int_\epsilon^R f(re^{i\varphi})e^{(a-1)i\varphi}e^{i\varphi}r^{a-1}\, dr$$

$$\to \int_\epsilon^R f(x)x^{a-1}\, dx \qquad \text{as} \quad \varphi \to 0.$$

On the other hand, $-L^-$ is parametrized by

$$z(r) = re^{i(2\pi - \varphi)}, \qquad \epsilon \leq r \leq R,$$

and $\log z(r) = \log r + i(2\pi - \varphi)$. Consequently,

$$\int_{L^-} f(z)e^{(a-1)\log z}\, dz = -\int_\epsilon^R f(re^{-i\varphi})r^{a-1}e^{(a-1)(2\pi - i\varphi)}e^{i(2\pi - i\varphi)}\, dr$$

$$= -\int_\epsilon^R f(re^{i\varphi})r^{a-1}e^{ai(2\pi - \varphi)}\, dr$$

$$\to -\int_\epsilon^r f(x)x^{a-1}e^{2\pi ia}\, dx \qquad \text{as} \quad \varphi \to 0.$$

Hence as $\varphi \to 0$, we find

$$\int_{L^+} + \int_{L^-} f(z)z^{a-1}\, dz \to \int_\epsilon^R f(x)x^{a-1}(1 - e^{2\pi ia})\, dx$$

$$= e^{\pi ia}\int_\epsilon^R f(x)x^{a-1}(e^{-\pi ia} - e^{\pi ia})\, dx$$

$$= 2ie^{\pi ia}\sin \pi a \int_\epsilon^R f(x)x^{a-1}\, dx.$$

Let $C = C(R, \epsilon, \varphi)$ denote the path of integration. We obtain

$$\int_{C(R,\,\epsilon,\,\varphi)} f(z)z^{a-1}\, dz = 2\pi i \sum \text{residues of } f(z)z^{a-1} \text{ except at } 0$$

$$= I_{S(R,\,\varphi)} + I_{S(\epsilon,\,\varphi)} + E(R,\,\epsilon,\,\varphi)$$

$$+ 2ie^{\pi ia}\sin \pi a \int_\epsilon^R f(x)x^{a-1}\, dx.$$

The expression $E(R, \epsilon, \varphi)$ denotes a term which goes to 0 as φ goes to 0, and we have put subscripts on the integrals along the arcs of the circle to show that they depend on R, ϵ, φ. We divide by $2ie^{\pi ia}\sin \pi a$, and let φ tend to 0. Then $E(R, \epsilon, \varphi)$ approaches 0. Consequently,

$$\int_\epsilon^R f(x)x^{a-1}\, dx - \frac{\pi e^{-\pi ia}}{\sin \pi a}\sum = \lim_{\varphi \to 0} \frac{I_{S(R,\,\varphi)} + I_{S(\epsilon,\,\varphi)}}{2ie^{\pi ia}\sin \pi a}.$$

The right-hand side has been seen to be uniformly small, independently of φ, and tends to 0 when $R \to \infty$ and $\epsilon \to 0$. Taking the limits as $R \to \infty$ and $\epsilon \to 0$ proves what we wanted.

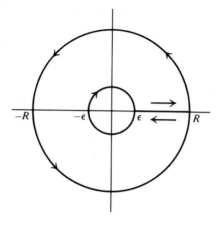

Figure 11

Finally, we observe that in situations of contour integrals as we just considered, it is often the practice to draw the limit contour as in Fig. 11. It is then understood that the value for the log when integrating over the segment from ϵ to R from left to right, and the value for the log when integrating over the segment from R to ϵ, are different, arising from the analytic expressions for the log with values $\theta = 0$ for the first and $\theta = 2\pi$ for the second.

For the Mellin transform of the gamma function, which provides an interesting special concrete case of the considerations of this section, see Exercise 6 of Chapter XII, §2.

EXERCISES VI §2

Find the following integrals.

1. $\displaystyle\int_{-\infty}^{\infty} \frac{1}{x^6 + 1}\, dx = 2\pi/3$

2. Show that for a positive integer n,

$$\int_0^{\infty} \frac{1}{1 + x^n}\, dx = \frac{\pi/n}{\sin \pi/n}.$$

[*Hint*: Try the path from 0 to R, then R to $Re^{2\pi i/n}$, then back to 0, or apply a general theorem.]

3. (a) $\displaystyle\int_{-\infty}^{\infty} \frac{x^2}{x^4 + 1}\, dx = \pi\sqrt{2}/2$ (b) $\displaystyle\int_0^{\infty} \frac{x^2}{x^6 + 1}\, dx = \pi/6$

4. Show that

$$\int_{-\infty}^{\infty} \frac{x-1}{x^5 - 1} \, dx = \frac{4\pi}{5} \sin \frac{2\pi}{5} \, .$$

5. Evaluate

$$\int_{\gamma} \frac{e^{-z^2}}{z^2} \, dz,$$

where γ is:
(a)   The square with vertices $1 + i$, $-1 + i$, $-1 - i$, $1 - i$.
(b)   The ellipse defined by the equation

$$\frac{x^2}{a^2} + \frac{y^2}{b^2} = 1.$$

(The answer is 0 in both cases.)

6. $\displaystyle\int_{-\infty}^{\infty} \frac{e^{iax}}{x^2 + 1} \, dx = \pi e^{-a}$ if $a > 0$

7. For any real number $a > 0$,

$$\int_{-\infty}^{\infty} \frac{\cos x}{x^2 + a^2} \, dx = \pi e^{-a}/a.$$

[*Hint*: This is the real part of the integral obtained by replacing $\cos x$ by e^{ix}.]

8. (a)   Show that for $a > 0$ we have

$$\int_{-\infty}^{\infty} \frac{\cos x}{(x^2 + a^2)^2} \, dx = \frac{\pi(1 + a)}{2a^3 e^a}.$$

(b)   Show that for $a > b > 0$ we have

$$\int_{0}^{\infty} \frac{\cos x}{(x^2 + a^2)(x^2 + b^2)} \, dx = \frac{\pi}{a^2 - b^2} \left(\frac{1}{be^b} - \frac{1}{ae^a} \right).$$

9. $\displaystyle\int_{0}^{\infty} \frac{\sin^2 x}{x^2} \, dx = \pi/2$ [*Hint*: Consider the integral of $(1 - e^{2ix})/x^2$.]

10. $\displaystyle\int_{-\infty}^{\infty} \frac{\cos x}{a^2 - x^2} \, dx = \frac{\pi \sin a}{a}$ for $a > 0$.   The integral is meant to be interpreted

as the limit:

$$\lim_{B \to \infty} \lim_{\delta \to 0} \int_{-B}^{-a-\delta} + \int_{-a+\delta}^{a-\delta} + \int_{a+\delta}^{B} .$$

11. $\displaystyle\int_{-\infty}^{\infty} \frac{\cos x}{e^x + e^{-x}}\, dx = \frac{\pi}{e^{\pi/2} + e^{-\pi/2}}.$ Use the indicated contour:

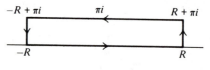

Figure 12

12. $\displaystyle\int_{0}^{\infty} \frac{x \sin x}{x^2 + a^2}\, dx = \frac{1}{2}\pi e^{-a}$ if $a > 0$.

13. $\displaystyle\int_{-\infty}^{\infty} \frac{e^{ax}}{e^x + 1}\, dx = \frac{\pi}{\sin \pi a}$ for $0 < a < 1$.

14. (a) $\displaystyle\int_{0}^{\infty} \frac{(\log x)^2}{1 + x^2}\, dx = \pi^3/8.$ Use the contour

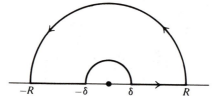

Figure 13

(b) $\displaystyle\int_{0}^{\infty} \frac{\log x}{(x^2 + 1)^2}\, dx = -\pi/4.$

15. (a) $\displaystyle\int_{0}^{\infty} \frac{x^a}{1 + x}\frac{dx}{x} = \frac{\pi}{\sin \pi a}$ for $0 < a < 1$.

(b) $\displaystyle\int_{0}^{\infty} \frac{x^a}{1 + x^3}\frac{dx}{x} = \frac{\pi}{3 \sin(\pi a/3)}$ for $0 < a < 3$.

16. Let f be a continuous function, and suppose that the integral

$$\int_{0}^{\infty} f(x)x^a \frac{dx}{x}$$

is absolutely convergent. Show that it is equal to the integral

$$\int_{-\infty}^{\infty} f(e^t)e^{at}\, dt.$$

If we put $g(t) = f(e^t)$, this shows that a Mellin transform is essentially a Fourier transform, up to a change of variable.

17. $\displaystyle\int_0^{2\pi} \frac{1}{1 + a^2 - 2a \cos \theta} \, d\theta = \frac{2\pi}{1 - a^2}$ if $0 < a < 1$. The answer comes out the negative of that if $a > 1$.

18. $\displaystyle\int_0^{\pi} \frac{1}{1 + \sin^2 \theta} \, d\theta = \pi/\sqrt{2}.$

19. $\displaystyle\int_0^{\pi} \frac{1}{3 + 2 \cos \theta} \, d\theta = \pi/\sqrt{5}.$

20. $\displaystyle\int_0^{\pi} \frac{a\,d\theta}{a^2 + \sin^2 \theta} = \int_0^{2\pi} \frac{a\,d\theta}{1 + 2a^2 - \cos \theta} = \frac{\pi}{\sqrt{1 + a^2}}.$

21. $\displaystyle\int_0^{\pi/2} \frac{1}{(a + \sin^2 \theta)^2} \, d\theta = \frac{\pi(2a + 1)}{4(a^2 + a)^{3/2}}$ for $a > 0$.

22. $\displaystyle\int_0^{2\pi} \frac{1}{2 - \sin \theta} \, d\theta = 2\pi/\sqrt{3}.$

23. $\displaystyle\int_0^{2\pi} \frac{1}{(a + b \cos \theta)^2} \, d\theta = \frac{2\pi a}{(a^2 - b^2)^{3/2}}$ for $0 < b < a$.

CHAPTER VII

Conformal Mappings

In this chapter we consider a more global aspect of analytic functions, describing geometrically what their effect is on various regions. Especially important are the analytic isomorphisms and automorphisms of various regions, of which we consider many examples.

Throughout the chapter, we use the words isomorphisms and automorphisms, omitting the word analytic, as there will be no others under consideration. We recall that an **isomorphism**

$$f: U \to V$$

is an analytic map which has an inverse analytic map

$$g: V \to U,$$

that is, $f \circ g = \mathrm{id}_V$ and $g \circ f = \mathrm{id}_U$. We say that f is an **automorphism** if $U = V$.

The main general theorem concerning isomorphisms is the **Riemann mapping theorem**:

If U is a simply connected open set which is not the whole plane, then there exists an isomorphism of U with the unit disc.

The general proof will be postponed to a later chapter. In the present chapter, we are concerned with specific examples, where the mapping can be exhibited concretely, in a simple manner.

It will also be useful to the reader to recall some simple algebraic formalism about isomorphisms and automorphisms, listed in the following properties.

Let $f: U \to V$ and $g: V \to W$ be two isomorphisms. Then

$$g \circ f : U \to W$$

is an isomorphism.

Let $f, g: U \to V$ be isomorphisms. Then there exists an automorphism h of V such that $g = h \circ f$.

Let $f: U \to V$ be an isomorphism. Then there is a bijection

$$\text{Aut}(U) \to \text{Aut}(V)$$

given by

$$\varphi \mapsto f \circ \varphi \circ f^{-1}.$$

The proofs are immediate in all cases. For instance, an inverse for $g \circ f$ is given by $f^{-1} \circ g^{-1}$ as one sees at once by composing these two maps in either direction. For the second statement, we have $h = g \circ f^{-1}$. As to the third, if φ is an automorphism of U then $f \circ \varphi \circ f^{-1}$ is an automorphism of V, because it is an isomorphism of V with itself. Similarly, if ψ is an automorphism of V then $f^{-1} \circ \psi \circ f$ is an automorphism of U. The reader will then verify that the associations

$$\varphi \mapsto f \circ \varphi \circ f^{-1} \quad \text{and} \quad \psi \mapsto f^{-1} \circ \psi \circ f$$

give maps between $\text{Aut}(U)$ and $\text{Aut}(V)$ which are inverse to each other, and hence establish the stated bijection between $\text{Aut}(U)$ and $\text{Aut}(V)$. The association $\varphi \mapsto f \circ \varphi \circ f^{-1}$ is called **conjugation by** f. It shows that if we know the set of automorphisms of U, then we also know the set of automorphisms of V if V is isomorphic to U: It is obtained by conjugation.

We also recall a result from Chapter II, Theorem 5.4.

Let f be analytic on an open set U. If f is injective, and $V = f(U)$ is its image, then

$$f : U \to V$$

is an analytic isomorphism, and in particular, $f'(z) \neq 0$ for all z in U.

This result came from the decomposition $f(z) = \varphi(z)^m$, where φ is a local analytic isomorphism in the neighborhood of a point z_0 in U, cf. Theorem 5.4.

Note that if f is analytic and $f'(z) \neq 0$ for all z in U, then we cannot conclude that f is injective. For instance let U be the open set obtained by deleting the origin from the plane, and let $f(z) = z^3$. Then $f'(z) \neq 0$ for all z in U, but f is not injective.

On the other hand, by restricting the open set U suitably, the map $z \mapsto z^3$ does become an isomorphism. For instance, let U be the sector consisting of all complex numbers $z = re^{i\theta}$ with $r > 0$ and $0 < \theta < \pi/3$. Then $z \mapsto z^3$ is an analytic isomorphism on U. What is its image?

VII §1. Schwarz Lemma

Let D be the unit disc of complex numbers z with $|z| < 1$.

Theorem 1.1. *Let $f: D \to D$ be an analytic function of the unit disc into itself such that $f(0) = 0$. Then:*

(i)   *We have $|f(z)| \leq |z|$ for all $z \in D$.*
(ii) *If for some $z_0 \neq 0$ we have $|f(z_0)| = |z_0|$, then there is some complex number α of absolute value 1 such that*

$$f(z) = \alpha z.$$

Proof. Let

$$f(z) = a_1 z + \cdots$$

be the power series for f. The constant term is 0 because we assumed $f(0) = 0$. Then $f(z)/z$ is holomorphic, and

$$\left| \frac{f(z)}{z} \right| < 1/r \qquad \text{for} \quad |z| = r < 1,$$

consequently also for $|z| \leq r$ by the maximum modulus principle. Letting r tend to 1 proves the first assertion. If furthermore we have

$$\left| \frac{f(z_0)}{z_0} \right| = 1$$

for some z_0 in the unit disc, then again by the maximum modulus principle $f(z)/z$ cannot have a maximum unless it is constant, and therefore there is a constant α such that $f(z)/z = \alpha$, whence the second assertion also follows.

In the above statement of the Schwarz lemma, the function was normalized to map the unit disc into itself. The lemma obviously implies analogous statements when the function satisfies a bound

$$|f(z)| \leq B \quad \text{on a disc} \quad |z| < R, \qquad \text{and} \qquad f(0) = 0.$$

The conclusion is then that

$$|f(z)| \leq B|z|/R,$$

and equality occurs at some point only if $f(z) = \dfrac{B}{R} \alpha z$, where α is a complex number of absolute value 1.

The following statement dealing with $f'(0)$ rather than the function itself will be considered as part of the Schwarz lemma.

Theorem 1.2. *Let $f: D \to D$ be an analytic function of the unit disc into itself such that $f(0) = 0$. Let*

$$f(z) = a_1 z + \text{higher terms}.$$

Then $|f'(0)| = |a_1| \leq 1$, and if $|a_1| = 1$, then $f(z) = a_1 z$.

Proof. Since $f(0) = 0$, the function $f(z)/z$ is analytic at $z = 0$, and

$$\frac{f(z)}{z} = a_1 + \text{higher terms}.$$

Letting z approach 0 and using the first part of Theorem 1.1 shows that $|a_1| \leq 1$. Next suppose $|a_1| = 1$ and

$$f(z) = a_1 z + a_m z^m + \text{higher terms}$$

with $a_m \neq 0$ and $m \geq 2$. Then

$$\frac{f(z)}{z} = a_1 + a_m z^{m-1} + \text{higher terms}.$$

Pick a value of z such that $a_m z^{m-1} = a_1$. There is a real number $C > 0$ such that for all small real $t > 0$ we have

$$\frac{f(tz)}{tz} = a_1 + a_m t^{m-1} z^{m-1} + h,$$

where $|h| \leq Ct^m$. Since $|a_1| = 1$, it follows that

$$\left| \frac{f(tz)}{tz} \right| = |a_1(1 + t^{m-1}) + h| > 1$$

for t sufficiently small. This contradicts the first part of Theorem 1.1, and concludes the proof.

VII §2. Analytic Automorphisms of the Disc

As an application of the Schwarz lemma, we shall determine all analytic automorphisms of the disc. First we give examples of such functions.

To begin with, we note that if φ is real, the map

$$z \mapsto e^{i\varphi}z$$

is interpreted geometrically as rotation counterclockwise by an angle φ. Indeed, if $z = re^{i\theta}$, then

$$e^{i\varphi}z = re^{i(\theta + \varphi)}.$$

Thus for example, the map $z \mapsto iz$ is a counterclockwise rotation by 90° (that is, $\pi/2$).

Let α be a complex number with $|\alpha| < 1$, and let

$$g_\alpha(z) = g(z) = \frac{\alpha - z}{1 - \bar{\alpha}z}.$$

Then g is analytic on the closed disc $|z| \leq 1$. Furthermore, if $|z| = 1$, then $z = e^{i\theta}$ for some real θ, and

$$g(z) = \frac{\alpha - e^{i\theta}}{e^{i\theta}(e^{-i\theta} - \bar{\alpha})}.$$

Up to the factor $e^{i\theta}$ which has absolute value 1, the denominator is equal to the complex conjugate of the numerator, and hence

$$if \ |z| = 1 \ then \ |g(z)| = 1.$$

By Theorem 1.2, if $|z| < 1$, then $|g(z)| < 1$. On the other hand, we can also argue now by the maximum modulus principle, that if $|z| \leq 1$, then $|g(z)| \leq 1$. By the open mapping theorem, it follows that if $|z| < 1$ then $|g(z)| < 1$. Furthermore, $g = g_\alpha$ has an inverse function. If we put $w = g_\alpha(z)$, then there is some β with $|\beta| < 1$ such that $z = g_\beta(w)$. As a trivial exercise, show that $\beta = \alpha$. Thus g_α and g_β are inverse functions on the unit disc, and thus $g = g_\alpha$ gives an analytic automorphism of the unit disc with itself.

Observe that $g(\alpha) = 0$. We now prove that up to rotations there are no other automorphisms of the unit disc.

Theorem 2.1. *Let $f: D \to D$ be an analytic automorphism of the unit disc and suppose $f(\alpha) = 0$. Then there exists a real number φ such that*

$$f(z) = e^{i\varphi} \frac{\alpha - z}{1 - \bar{\alpha} z}.$$

Proof. Let $g = g_\alpha$ be the above automorphism. Then $f \circ g^{-1}$ is an automorphism of the unit disc, and maps 0 on 0, i.e. it has a zero at 0. It now suffices to prove that the function $h(w) = f(g^{-1}(w))$ is of the form

$$h(w) = e^{i\varphi} w$$

to conclude the proof of the theorem.

The first part of Schwarz lemma tells us that

$$|h(z)| \leq |z| \quad \text{if} \quad |z| < 1.$$

Since the inverse function h^{-1} also has a zero at the origin, we also get the inequality in the opposite direction, that is,

$$|z| \leq |h(z)|,$$

and the second part of Schwarz lemma now implies that $h(z) = e^{i\varphi} z$, thereby proving our theorem.

Corollary 2.2. *If f is an automorphism of the disc which leaves the origin fixed, i.e. $f(0) = 0$, then $f(z) = e^{i\varphi} z$ for some real number φ, so f is a rotation.*

Proof. Let $\alpha = 0$ in the theorem.

EXERCISES VII §2

1. Let f be analytic on the unit disc D, and assume that $|f(z)| < 1$ on the disc. Prove that if there exist two distinct points a, b in the disc which are fixed points, that is, $f(a) = a$ and $f(b) = b$, then $f(z) = z$.

2. **(Schwarz–Pick Lemma).** Let $f: D \to D$ be a holomorphic map of the disc into itself. Prove that for all $a \in D$ we have

$$\frac{|f'(a)|}{1 - |f(a)|^2} \leq \frac{|a|}{1 - |a|^2}.$$

[*Hint*: Let g be an automorphism of D such that $g(0) = a$, and let h be an automorphism which maps $f(a)$ on 0. Let $F = h \circ f \circ g$. Compute $F'(0)$ and apply the Schwarz lemma.]

3. Let α be a complex number, and let h be an isomorphism of the disc $D(\alpha, R)$ with the unit disc such that $h(z_0) = 0$. Show that

$$h(z) = \frac{R(z - z_0)}{R^2 - (z - \alpha)(\bar{z}_0 - \bar{\alpha})} e^{i\theta}$$

for some real number θ.

4. What is the image of the half strips as shown on the figure, under the mapping $z \mapsto iz$? Under the mapping $z \mapsto -iz$?

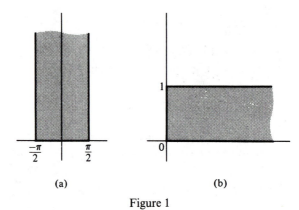

(a)   (b)

Figure 1

5. Let α be real, $0 \leqq \alpha < 1$. Let U_α be the open set obtained from the unit disc by deleting the segment $[\alpha, 1]$, as shown on the figure.

(a) Find an isomorphism of U_α with the unit disc from which the segment $[0, 1]$ has been deleted.

(b) Find an isomorphism of U_0 with the upper half of the disc. Also find an isomorphism of U_α with this upper half disc.

[*Hint*: What does $z \mapsto z^2$ do to the upper half disc?]

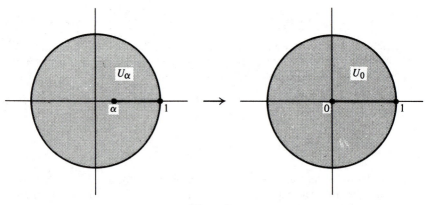

Figure 2

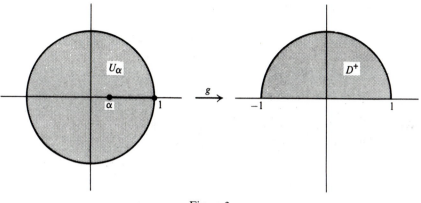

Figure 3

VII §3. The Upper Half Plane

Theorem 3.1. *Let H be the upper half plane. The map*

$$f : z \mapsto \frac{z - i}{z + i}$$

is an isomorphism of H with the unit disc.

Proof. Let $w = f(z)$ and $z = x + iy$. Then

$$f(z) = \frac{x + (y - 1)i}{x + (y + 1)i}.$$

Since z is in H, $y > 0$, it follows that $(y - 1)^2 < (y + 1)^2$ whence

$$x^2 + (y - 1)^2 = |z - i|^2 < x^2 + (y + 1)^2 = |z + i|^2$$

and therefore

$$|z - i| < |z + i|,$$

so f maps the upper half plane into the unit disc. Since

$$w = \frac{z - i}{z + i},$$

we can solve for z in terms of w, because $wz + wi = z - i$, so that

$$z = -i\frac{w+1}{w-1}.$$

Write $w = u + iv$, with real u, v. By computing directly the real part of $(w + 1)/(w - 1)$, and so the imaginary part of

$$-i\frac{w+1}{w-1}$$

you will find that this imaginary part is >0 if $|w| < 1$. Hence the map

$$h: w \mapsto -i\frac{w+1}{w-1}$$

sends the unit disc into the upper half plane. Since by construction f and h are inverse to each other, it follows that they are inverse isomorphisms of the upper half plane and the disc, as was to be shown.

Example. We wish to give an isomorphism of the first quadrant with the unit disc. Since we know that the upper half plane is isomorphic to the unit disc, it suffices to exhibit an isomorphism of the first quadrant with the upper half plane. The map

$$z \mapsto z^2$$

achieves this.

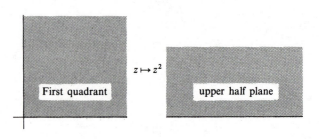

Figure 4

If $f: H \to D$ is the isomorphism of the upper half plane with the unit disc then

$$z \mapsto f(z^2)$$

is the desired isomorphism of the first quadrant with the unit disc. Thus the function

$$z \mapsto \frac{z^2 - i}{z^2 + i}$$

gives an isomorphism of the first quadrant with the unit disc.

EXERCISES VII §3

Let

$$\gamma = \begin{pmatrix} a & b \\ c & d \end{pmatrix}$$

be a 2×2 matrix of real numbers, such that $ad - bc > 0$. For $z \in H$, define

$$f_\gamma(z) = \frac{az + b}{cz + d}.$$

1. Show that

$$\operatorname{Im} f_\gamma(z) = \frac{(ad - bc)y}{|cz + d|^2}.$$

2. Show that f_γ gives a map of H into H.

3. Show that there exists a matrix

$$\gamma' = \begin{pmatrix} a' & b' \\ c' & d' \end{pmatrix}$$

of real numbers with $a'b' - c'd' > 0$ such that $f_{\gamma'}$ is an inverse of f, so f is an isomorphism of H with itself.

It can be shown by brute force using Theorem 2.1 that the automorphisms of type f_γ give all possible automorphisms of H.

4. Let $f(z) = e^{2\pi i z}$. Show that f maps the upper half plane on the inside of a disc from which the center has been deleted. Given $B > 0$, let $H(B)$ be that part of the upper half plane consisting of those complex numbers $z = x + iy$ with $y \geq B$. What is the image of $H(B)$ under f? Is f an isomorphism? Why? How would you restrict the domain of definition of f to make it an isomorphism?

VII §4. Other Examples

We give the examples by pictures which illustrate various isomorphisms.

Example 1.

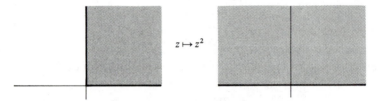

$$z \mapsto z^2$$

Isomorphism between first quadrant and upper half plane

Figure 5

Example 2.

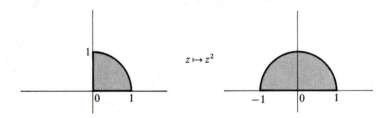

$$z \mapsto z^2$$

Isomorphism between quarter disc and half disc

Figure 6

Example 3.

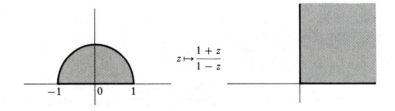

$$z \mapsto \frac{1+z}{1-z}$$

Upper half disc with first quadrant

Figure 7

Example 4. By composing the above isomorphisms, we get new ones. For instance, let U be the portion of the unit disc lying inside the first quadrant as in Example 2. We want to get an isomorphism of U with the upper half plane.

All we have to do is to compose the isomorphisms of Examples 2, 3, and 1 in that order. Thus an isomorphism of U with H is given by the formula in the picture.

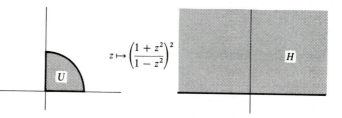

Quarter disc with upper half plane

Figure 8

The next three examples concern the logarithm.

Example 5.

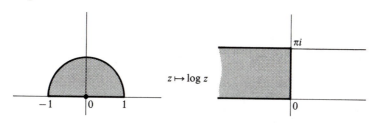

Upper half disc with a half strip

Figure 9

Example 6.

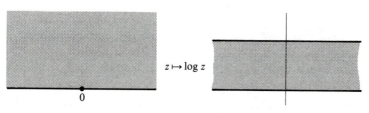

Upper half plane with a full strip

Figure 10

Example 7.

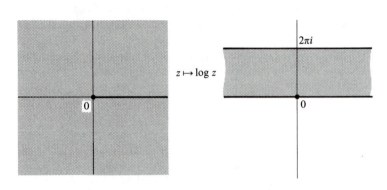

Plane with deleted positive real axis with a full strip

Figure 11

In the applications to fluid dynamics, we shall see in the next chapter that it is important to get isomorphisms of various regions with the upper half plane in order to be able to describe the flow lines. In particular, certain regions are obtained by placing obstacles inside simpler regions. We give several examples of this phenomenon. These will allow us to get an isomorphism from a strip containing a vertical obstacle with the upper half plane.

Example 8.

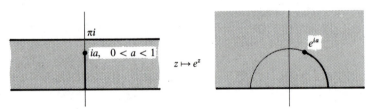

Figure 12

Example 9.

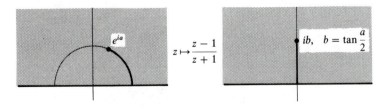

Figure 13

Example 10.

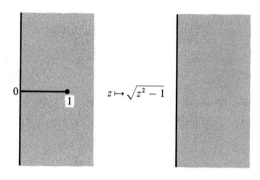

Figure 14

Remark. By composing the isomorphisms of Examples 8 and 9, using a dilation, and a rotation, and finally the isomorphism of Example 10, we get an isomorphism of the strip containing a vertical obstacle with the right half plane:

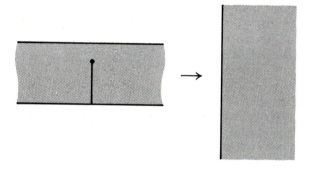

Figure 15

Another rotation would then yield the upper half plane.

Example 11. In this example, the obstacle is a bump rather than a vertical line segment. *We claim that the map*

$$z \mapsto z + \frac{1}{z}$$

is an isomorphism of the open set U lying inside the upper half plane, above the unit circle, with the upper half plane.

The isomorphism is shown on the following figure.

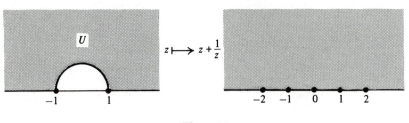

Figure 16

Proof. Let $w = z + 1/z$ so that

$$w = x\left(1 + \frac{1}{x^2 + y^2}\right) + iy\left(1 - \frac{1}{x^2 + y^2}\right).$$

If $z \in U$, then $|z| > 1$ so Im $w > 0$ and $w \in H$. The quadratic equation

$$z^2 - zw + 1 = 0$$

has two distinct roots except for $w = \pm 2$.   Given $w \in H$, and root $z = x + iy$ has the property that either $y > 0$ and $x^2 + y^2 > 1$, or $y < 0$ and $x^2 + y^2 < 1$. Since the product of the two roots is 1 (from the quadratic equation), and hence the product of their absolute values is also 1, it follows that not both roots can have absolute value > 1 or both have absolute value < 1. Hence exactly one root lies in U, so the map is both surjective and injective, as desired.

Example 12.

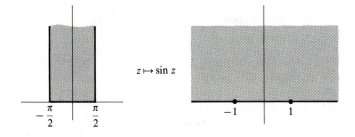

Upper half strip with upper half plane

Figure 17

The sine function maps the interval $[-\pi/2, \pi/2]$ on the interval $[-1, 1]$.

Let us look also at what the sine does to the right vertical boundary, which consists of all points $\pi/2 + it$ with $t \geq 0$. We know that

$$\sin z = \frac{e^{iz} - e^{-iz}}{2i}.$$

Hence

$$\sin\left(\frac{\pi}{2} + it\right) = \frac{e^{i\pi/2}e^{iit} - e^{-i\pi/2}e^{-iit}}{2i}$$

$$= \frac{e^t + e^{-t}}{2}.$$

As t ranges from 0 to infinity, $\sin(\pi/2 + it)$ therefore ranges from 1 to infinity, so the image of the vertical half line is the part of the real axis lying to the right of 1. You can show similarly that the image of the left vertical boundary is the part of the real axis to the left of -1. Thus we see precisely what the mapping $z \mapsto \sin z$ does to the boundary of the region.

For the convenience of the reader, we also discuss the mapping on the interior of the region. Let $z = x + iy$ with

$$-\frac{\pi}{2} \leq x \leq \frac{\pi}{2} \quad \text{and} \quad 0 < y.$$

From the definition of $\sin z$, letting $w = \sin z = u + iv$, we find

(1) $u = \sin x \cosh y \quad \text{and} \quad v = \cos x \sinh y,$

where

$$\cosh y = \frac{e^y + e^{-y}}{2} \quad \text{and} \quad \sinh y = \frac{e^y - e^{-y}}{2}.$$

From (1) we get

(2) $$\frac{u^2}{\cosh^2 y} + \frac{v^2}{\sinh^2 y} = 1,$$

(3) $$\frac{u^2}{\sin^2 x} - \frac{v^2}{\cos^2 x} = 1.$$

If we fix a value of $y > 0$, then the line segment

$$x + iy \quad \text{with} \quad -\frac{\pi}{2} \leq x \leq \frac{\pi}{2}$$

gets mapped onto the upper half of an ellipse in the w-plane, as shown on the figure. Note that for the given intervals, we have $u \geq 0$.

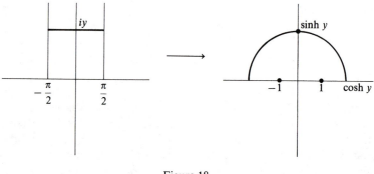

Figure 18

Geometrically speaking, as y increases from 0 to infinity, the ellipses expand and fill out the upper half plane.

One can also determine the image of vertical lines, fixing x and letting y vary, so half lines of the form $x + iy$ with $y > 0$ and x fixed. Equation (3) shows that the image of such half lines are upper parts of hyperbolas. It is the right upper part if $x > 0$ and the left upper part if $x < 0$. Since an analytic map with non-zero derivative is conformal, these hyperbolas are perpendicular to the above ellipses because vertical lines are perpendicular to horizontal lines.

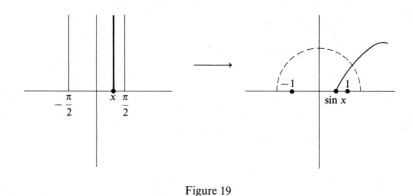

Figure 19

EXERCISES VII §4

1. (a) In each one of the examples, prove that the stated mapping is an isomorphism on the figures as shown. Also determine what the mapping does to the boundary lines. Thick lines should correspond to each other.
   (b) In Example 10, give the explicit formula giving an isomorphism of the strip containing a vertical obstacle with the right half plane, and also with the upper half plane. Note that counterclockwise rotation by $\pi/2$ is given by multiplication with i.

2. (a) Show that the function $z \mapsto z + 1/z$ is an analytic isomorphism of the region outside the unit circle onto the plane from which the segment $[-2, 2]$ has been deleted.
   (b) What is the image of the unit circle under this mapping? Use polar coordinates.

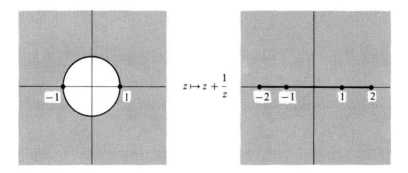

Figure 20

(c) In polar coordinates, if $w = z + 1/z = u + iv$, then

$$u = \left(r + \frac{1}{r}\right)\cos\theta \quad \text{and} \quad v = \left(r - \frac{1}{r}\right)\sin\theta.$$

Show that the circle $r = c$ with $c > 1$ maps to an ellipse with major axis $c + 1/c$ and minor axis $c - 1/c$. Show that the radial lines $\theta = c$ map onto quarters of hyperbolas.

3. Let a be a real number. Let U be the open set obtained from the complex plane by deleting the infinite segment $[a, \infty[$. Find explicitly an analytic isomorphism of U with the unit disc. Give this isomorphism as a composite of simpler ones. [*Hint*: Try first to see what $\sqrt{z}$ does to the set obtained by deleting $[0, \infty[$ from the plane.]

4. (a) Show that the function $w = \sin z$ can be decomposed as the composite of two functions:

$$w = \frac{\zeta + \zeta^{-1}}{2i} = f(\zeta) \quad \text{and} \quad \zeta = e^{iz} = g(z).$$

(b) Let U be the open upper half strip in Example 12. Let $g(U) = V$. Describe V explicitly and show that $g: U \to V$ is an isomorphism. Show that g extends to a continuous function on the boundary of U and describe explicitly the image of this boundary under g.

(c) Let $W = f(V)$. Describe W explicitly and show that $f: V \to W$ is an isomorphism. Again describe how f extends continuously to the boundary of V and describe explicitly the image of this boundary under f.

In this way you can recover the fact that $w = \sin z$ gives an isomorphism of the upper half strip with the upper half plane by using this decomposition into simpler functions which you have already studied.

5. In Example 12, show that the vertical imaginary axis above the real line is mapped onto itself by $z \mapsto \sin z$, and that this function gives an isomorphism of the half strip with the first quadrant as shown on the figure.

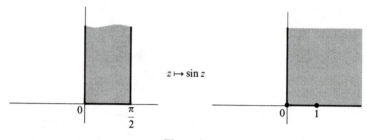

Figure 21

6. Let $w = u + iv = f(z) = z + \log z$ for z in the upper half plane H. Prove that f gives an isomorphism of H with the open set U obtained from the upper half plane by deleting the infinite half line of numbers

$$u + i\pi \qquad \text{with} \quad u \leqq -1.$$

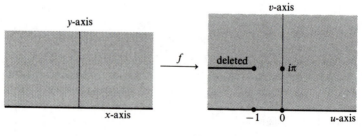

Figure 22

Remark. In the next chapter, we shall see that the isomorphism f allows us to determine the flow lines of a fluid as shown on the figure. These flow lines in the (u, v)-plane correspond to the horizontal lines $y = \text{constant}$ in the (x, y)-plane. In other words, they are the images under f of the horizontal lines $y = \text{constant}$.

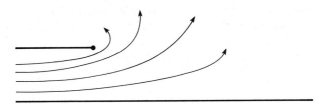

Flow lines in the (u, v)-plane

Figure 23

[*Hint*: Let $w = u + iv$, so that

$$u = x + \log r = r \cos \theta + \log r,$$

$$v = y + \theta \quad r = \quad \sin \theta + \theta.$$

Note that $f'(z) \neq 0$ for all z. First show that f is injective. Then show that f is surjective as follows. Fix a value of $v > 0$. Thus

$$r = \frac{v - \theta}{\sin \theta} \quad \text{with} \quad 0 < \theta < v,$$

and

$$u = (v - \theta) \cot \theta + \log \frac{v - \theta}{\sin \theta}.$$

Distinguish cases $v > \pi$, $v < \pi$, and $v = \pi$. In each case with $v \neq \pi$, show that u takes on all values as θ ranges from 0 to v. When $v = \pi$, show that the missing values are exactly $u \leq -1$. Look at what happens to u when θ is near 0, and when θ is near v. Use the intermediate value theorem from calculus to get all intermediate values].

VII §5. Fractional Linear Transformations

Let a, b, c, d be complex numbers such that $ad - bc \neq 0$. We may arrange these numbers as a matrix $\begin{pmatrix} a & b \\ b & d \end{pmatrix}$. Let

$$F(z) = \frac{az + b}{cz + d}.$$

We call F a **fractional linear map**, or **transformation**. We have already encountered functions of this type, and now we study them more systematically. First observe that if we multiply a, b, c, d by the same non-zero complex number λ, then the matrix

$$\begin{pmatrix} \lambda a & \lambda b \\ \lambda c & \lambda d \end{pmatrix}$$

gives rise to the same map, because we can cancel λ in the fraction:

$$\frac{\lambda a z + \lambda b}{\lambda c z + \lambda d} = \frac{az + b}{cz + d}.$$

It is an exercise to prove the converse, that if two matrices

$$\begin{pmatrix} a & b \\ c & d \end{pmatrix} \quad \text{and} \quad \begin{pmatrix} a' & b' \\ c' & d' \end{pmatrix}$$

of complex numbers with $ad - bc \neq 0$ and $a'd' - b'c' \neq 0$ give the same fractional linear map, then there is a complex number λ such that

$$a' = \lambda a, \qquad b' = \lambda b, \qquad c' = \lambda c, \qquad d' = \lambda d.$$

We shall now see that F gives an isomorphism. Note that

$$F'(z) = \frac{ad - bc}{(cz + d)^2}.$$

The function F is not defined at $z = -d/c$, but is defined at all other complex numbers, and the formula for its derivative then shows that $F'(z) \neq 0$ for all complex numbers $z \neq -d/c$.

The function F has an inverse. Indeed, let

$$w = \frac{az + b}{cz + d}.$$

We can solve for z in terms of w by simple algebra. Cross multiplying yields

$$czw + dw = az + b,$$

whence

$$z = \frac{dw - b}{-cw + a}.$$

Thus the inverse function is associated with the matrix

$$\begin{pmatrix} d & -b \\ -c & a \end{pmatrix}.$$

Observe that the inverse function, which we denote by F^{-1}, is not defined at $z = a/c$. Thus F gives an isomorphism of $\mathbf{C}$ from which $-d/c$ has been deleted with $\mathbf{C}$ from which a/c has been deleted.

To have a uniform language to deal with the "exceptional" points $z = -d/c$ and $z = a/c$, we agree to the following conventions.

Let S be the Gauss sphere, i.e. the set consisting of $\mathbf{C}$ and a single point ∞ which we call **infinity**. We extend the definition of F to S by defining

$$F(\infty) = a/c \qquad \text{if} \quad c \neq 0,$$
$$F(\infty) = \infty \qquad \text{if} \quad c = 0.$$

Also we define

$$F(-d/c) = \infty.$$

These definitions are natural, for if we write

$$F(z) = \frac{a + b/z}{c + d/z},$$

and let $|z| \to \infty$ then this fraction approaches a/c as a limit.

We may then say that F gives a bijection of S with itself.

We now define other maps as follows:

$T_b(z) = z + b$, called **translation** by b;

$J(z) = 1/z$, called **inversion** through the unit circle;

$M_a(z) = az$ for $a \neq 0$, called **multiplication** by a.

Observe that translations, reflections, or multiplications are fractional linear maps. Translations should have been encountered many times previously. As for inversion, note:

If $|z| = 1$ then $1/z = \bar{z}$ and $|1/z| = 1$ also. Thus an inversion maps the unit circle onto itself.

If $|z| > 1$ then $|1/z| < 1$ and vice versa, so an inversion interchanges the region outside the unit disc with the region inside the unit disc. Note that 0 and ∞ correspond to each other under the inversion.

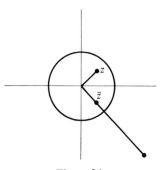

Figure 24

Multiplication by a complex number a can be viewed as a dilation together with a rotation, by writing $a = re^{i\theta}$.

Thus each one of these particular linear maps has a simple geometric interpretation as above.

Theorem 5.1. *Given a fractional linear map* F, *there exist complex numbers* α, β, γ *such that either* $F = \alpha z + \beta$, *or*

$$F(z) = T_\gamma \circ M_\alpha \circ J \circ T_\beta.$$

Proof. Suppose $c = 0$. Then $F(z) = (az + b)/d$ and $F = T_\beta \circ M_\alpha$, with $\beta = b/d$, $\alpha = a/d$. Suppose this is not the case, so $c \neq 0$. We divide a, b, c, d by c and using these new numbers gives the same map F, so without loss of generality we may assume $c = 1$. We let $\beta = d$. We must solve

$$\frac{az + b}{z + d} = \frac{\alpha}{z + d} + \gamma,$$

or in other words, $az + b = \alpha + \gamma z + \gamma d$. We let $\gamma = a$, and then solve for $\alpha = b - ad \neq 0$ to conclude the proof.

The theorem shows that any fractional linear map is a composition of the simple maps listed above: translation, multiplication, or inversion.

Now let us define a **straight line on the Riemann sphere** S to consist of an ordinary line together with ∞.

Theorem 5.2. *A fractional linear transformation maps straight lines and circles onto straight lines and circles.* (*Of course, a circle may be mapped onto a line and vice versa.*)

Proof. By Theorem 5.1 it suffices to prove the assertion in each of the three cases of the simple maps. The assertion is obvious for translations and multiplications (which are rotations followed by dilations). There remains to deal with the inversion. The equation of a circle or straight line can be written in the form

$$A(x^2 + y^2) + Bx + Cy = D$$

with some real numbers A, B, C, D such that not all A, B, C are equal to 0. Let $w = 1/z = u + iv$, so that

$$u = \frac{x}{x^2 + y^2} \quad \text{and} \quad v = \frac{-y}{x^2 + y^2}.$$

Then in terms of u, v the preceding equation has the form

$$-D(u^2 + v^2) + Bu - Cv = -A,$$

which is the equation of a circle of a straight line. This proves the theorem.

As an exercise, you will prove that if F, G are fractional linear maps, then so is $F \circ G$. We shall use such compositions in the next considerations.

By a **fixed point** of F we mean a point z_0 such that $F(z_0) = z_0$.

Example. The point ∞ is a fixed point of the map

$$F(z) = az + b.$$

Proposition 5.3. *Let F be fractional linear map. If ∞ is a fixed point of F, then there exist complex numbers a, b such that $F(z) = az + b$.*

Proof. Let $F(z) = (az + b)/(cz + d)$. If $c \neq 0$ then $F(\infty) = a/c$ which is not ∞. By hypothesis, it follows that $c = 0$, in which case the assertion is obvious.

Theorem 5.4. *Given any three distinct points z_1, z_2, z_3 on the Riemann sphere, and any other three distinct points w_1, w_2, w_3, there exists a unique fractional linear map F such that*

$$F(z_i) = w_i \qquad for \quad i = 1, 2, 3.$$

Proof. We proceed stepwise, and first prove uniqueness. Let F, G be fractional linear maps which have the same effect on three points. Then $F \circ G^{-1}$ has three fixed points, and it suffices to prove the following lemma.

Lemma 5.5. *Let F be a fractional linear map. If F has three fixed points, then F is the identity.*

Proof. Suppose first that one fixed point is ∞. By Proposition 5.3, we know that $F(z) = az + b$. If z_1, z_2 are two other fixed points, then we have

$$az_1 + b = z_1,$$

$$az_2 + b = z_2.$$

Subtracting shows that $a = 1$, and then $b = 0$, thus proving the lemma, and concluding the proof of uniqueness if one of the fixed points is ∞.

In general, given any point z_1 there exists a fractional linear map T such that $T(z_1) = \infty$ (proof?). Then $T \circ F \circ T^{-1}$ is a fractional linear map. If z_1, z_2, z_3 are fixed points of F, then

$$T(z_1), \; T(z_2), \; T(z_3)$$

are fixed points of $T \circ F \circ T^{-1}$, as one sees directly, because

$$T \circ F \circ T^{-1}(T(z)) = T(F(z)).$$

By the first part of the proof, and the fact that ∞ is a fixed point of $T \circ F \circ T^{-1}$, we conclude that $T \circ F \circ T^{-1} = \text{id}$, whence $F = \text{id}$. This proves the lemma and concludes the proof of uniqueness.

For the existence, it will suffice to prove that given three points z_1, z_2, z_3 there exists a fractional linear map F such that

$$F(z_1) = 0, \qquad F(z_2) = \infty, \qquad \text{and} \qquad F(z_3) = 1.$$

Indeed, if we can do this, then there is also a fractional linear map G such that

$$G(w_1) = 0, \qquad G(w_2) = \infty, \qquad \text{and} \qquad G(w_3) = 1,$$

and therefore $G^{-1} \circ F(z_i) = w_i$ for $i = 1, 2, 3$ thus proving Theorem 5.4.

So we first let T be a fractional linear map such that $T(z_2) = \infty$. Let $T(z_1) = z'_1$ and $T(z_3) = z'_3$. It suffices to find complex numbers $a \neq 0$ and b such that if we let $G(z) = az + b$, then $G(z'_1) = 0$ and $G(z'_3) = 1$, because then we let $F = G \circ T$. But we can solve the linear equations

$$az'_1 + b = 0,$$
$$az'_3 + b = 1$$

(do so explicitly) thus concluding the proof.

Remark. The reader will now appreciate the technique which consists in mapping one point to ∞. This makes computations very much easier, since we can then deal with the easier fractional linear maps of the form $az + b$ which leave ∞ fixed.

One can give an easy formula for the map F as above. Note that the function $z \mapsto z - z_1$ sends z_1 to 0. Then

$$z \mapsto \frac{z - z_1}{z - z_2}$$

sends z_1 to 0 and z_2 to ∞. To send z_3 to 1, all we have to do is multiply by the right factor, and thus we obtain:

Theorem 5.6. *The function*

$$z \mapsto \frac{z - z_1}{z - z_2} \frac{z_3 - z_2}{z_3 - z_1}$$

is the unique function such that $F(z_1) = 0$, $F(z_2) = \infty$, $F(z_3) = 1$. If $w = F(z)$ is the function such that $F(z_i) = w_i$ for $i = 1, 2, 3$ then w and z are related by the formula

$$\boxed{\frac{w - w_1}{w - w_2} \frac{w_3 - w_2}{w_3 - w_1} = \frac{z - z_1}{z - z_2} \frac{z_3 - z_2}{z_3 - z_1}.}$$

This final equation can be used to find F explicitly in special cases.

Example. Find the map F in Theorem 5.4 such that

$$F(1) = i, \qquad F(i) = -1, \qquad F(-1) = 1.$$

By the formula,

$$\frac{w - i}{w + 1} \frac{1 + 1}{1 - i} = \frac{z - 1}{z - i} \frac{-1 - i}{-1 - 1},$$

or in other words,

$$\frac{w - i}{w + 1} = \frac{1}{2} \frac{z - 1}{z - i}.$$

We can solve for w in terms of z to give

$$w = \frac{z(1 + 2i) + 1}{z + (1 - 2i)}.$$

To check the computation, substitute $z = 1$, $z = i$, $z = -1$ in this expression to see that you get the desired values i, -1, and 1, respectively.

Warning. I find it pointless to memorize the formula in Theorem 5.6 relating z and w. However, the comments before Theorem 5.6 tell you how to reconstruct this formula in an easy way if you don't have it for reference in front of you.

EXERCISES VII §5

1. Give explicitly a fractional linear map which sends a given complex number z_1 to ∞. What is the simplest such map which sends 0 to ∞?

2. **Composition of Fractional Linear Maps.** Show that if F, G are fractional linear maps, then so is $F \circ G$. If you know about the multiplication of matrices, and σ, σ' are two 2×2 matrices with non-zero determinant, show that

$$G_\sigma \circ G_{\sigma'} = G_{\sigma\sigma'}.$$

3. Find fractional linear maps which map:
   (a) $1, i, -1$ on $i, -1, 1$
   (b) $i, -1, 1$ on $-1, -i, 1$
   (c) $-1, -i, 1$ on $-1, 0, 1$
   (d) $-1, 0, 1$ on $-1, i, 1$
   (e) $-1, i,$ on $1, i, -1$

4. Find fractional linear maps which map:
   (a) $0, 1, \infty$ on $1, \infty, 0$
   (b) $0, 1, \infty$ on $1, -1, i$
   (c) $0, 1, \infty$ on $-1, 0, 1$
   (d) $0, 1, \infty$ on $-1, -i, 1$

5. Let F and G be two fractional linear maps, and assume that $F(z) = G(z)$ for all complex numbers z (or even for three distinct complex numbers z). Show that if

$$F(z) = \frac{az + b}{cz + d} \quad \text{and} \quad G(z) = \frac{a'z + b'}{c'z + d'}$$

then there exists a complex number λ such that

$$a' = \lambda a, \qquad b' = \lambda b, \qquad c' = \lambda c, \qquad d' = \lambda d.$$

Thus the matrices representing F and G differ by a scalar.

6. Consider the fractional linear map

$$F(z) = \frac{z - i}{z + i}.$$

What is the image of the real line **R** under this map? (You have encountered this map as an isomorphism between the upper half plane and the unit disc.)

7. Let F be the fractional linear map $F(z) = (z - 1)/(z + 1)$. What is the image of the real line under this map? (Cf. Example 9 of §4.)

8. Let $F(z) = z/(z - 1)$ and $G(z) = 1/(1 - z)$. Show that the set of all possible fractional linear maps which can be obtained by composing F and G above repeatedly with each other in all possible orders in fact has six elements, and give a formula for each one of these elements. [*Hint*: Compute F^2, F^3, G^2, G^3, $F \circ G$, $G \circ F$, etc.]

9. Let $F(z) = (z - i)/(z + i)$. What is the image under F of the following sets of points:
   (a) The upper half line it, with $t \geq 0$.
   (b) The circle of center 1 and radius 1.
   (c) The horizontal line $i + t$, with $t \in \mathbf{R}$.
   (d) The half circle $|z| = 2$ with $\mathrm{Im}\ z \geq 0$.
   (e) The vertical line $\mathrm{Re}\ z = 1$ and $\mathrm{Im}\ z \geq 0$.

10. Find fractional linear maps which map:
    (a) $0, 1, 2$ to $1, 0, \infty$
    (b) $i, -1, 1$ to $1, 0, \infty$
    (c) $0, 1, 2$ to $i, -1, 1$

11. Let $F(z) = (z + 1)/(z - 1)$. Describe the image of the line $\mathrm{Re}(z) = c$ for a real number c. (Distinguish $c = 1$ and $c \neq 1$. In the second case, the image is a circle. Give its center and radius.)

12. Let z_1, z_2, z_3, z_4 be distinct complex numbers. Define their **cross ratio** to be

$$[z_1, z_2, z_3, z_4] = \frac{(z_1 - z_3)(z_2 - z_4)}{(z_2 - z_3)(z_1 - z_4)}.$$

   (a) Let F be a fractional linear map. Let $z_i' = F(z_i)$ for $i = 1, \ldots, 4$. Show that the cross ratio of z_1', z_2', z_3', z_4' is the same as the cross ratio of z_1, z_2, z_3, z_4. It will be easy if you do this separately for translations, inversions, and multiplications.
   (b) Prove that the four numbers lie on the same straight line or on the same circle if and only if their cross ratio is a real number.
   (c) Let z_1, z_2, z_3, z_4 be distinct complex numbers. Assume that they lie on the same circle, in that order. Prove that

$$|z_1 - z_3||z_2 - z_4| = |z_1 - z_2||z_3 - z_4| + |z_2 - z_3||z_4 - z_1|.$$

CHAPTER VIII

Harmonic Functions

In this chapter we return to the connection between analytic functions and functions of a real variable, analyzing an analytic function in terms of its real part.

The first two sections, §1 and §2, are completely elementary and could have been covered in Chapter I. They combine well with the material in the preceding chapter, as they deal with the same matter, pursued to analyze the real part of analytic isomorphisms more closely.

In §3 and §4 we deal with those aspects of harmonic functions having to do with integration and some form of Cauchy's formula. We shall characterize harmonic functions as real parts of analytic functions, giving an explicit integral formula for the associated analytic function (uniquely determined except for a pure imaginary constant).

VIII §1. Definition

A function $u = u(x, y)$ is called **harmonic** if it is real valued having continuous partial derivatives of order one and two, and satisfying

$$\frac{\partial^2 u}{\partial x^2} + \frac{\partial^2 u}{\partial y^2} = 0.$$

Suppose f is analytic on an open set U. We know that f is infinitely complex differentiable. By the considerations of Chapter I, §6, it follows that its real and imaginary parts $u(x, y)$ and $v(x, y)$ are C^∞, and satisfy the Cauchy–Riemann equations

$$\frac{\partial u}{\partial x} = \frac{\partial v}{\partial y} \quad \text{and} \quad \frac{\partial u}{\partial y} = -\frac{\partial v}{\partial x}.$$

Consequently, taking the partial derivatives of these equations and using the known fact that $\dfrac{\partial}{\partial x}\dfrac{\partial}{\partial y} = \dfrac{\partial}{\partial y}\dfrac{\partial}{\partial x}$ yields:

Theorem 1.1. *The real part of an analytic function is harmonic.*

Example. Let $r = \sqrt{x^2 + y^2}$. Then $\log r$ is harmonic, being the real part of the complex log.

We introduce the differential operators

$$\frac{\partial}{\partial \bar{z}} = \frac{1}{2}\left(\frac{\partial}{\partial x} + i\frac{\partial}{\partial y}\right) \quad \text{and} \quad \frac{\partial}{\partial z} = \frac{1}{2}\left(\frac{\partial}{\partial x} - i\frac{\partial}{\partial y}\right).$$

The reason for this notation is apparent if we write

$$x = \frac{1}{2}(z + \bar{z}) \quad \text{and} \quad y = \frac{1}{2i}(z - \bar{z}).$$

We want the chain rule to hold. Working formally, we see that the following equations must be satisfied.

$$\frac{\partial f}{\partial z} = \frac{\partial f}{\partial x}\frac{\partial x}{\partial z} + \frac{\partial f}{\partial y}\frac{\partial y}{\partial z} = \frac{1}{2}\frac{\partial f}{\partial x} + \frac{1}{2i}\frac{\partial f}{\partial y},$$

$$\frac{\partial f}{\partial \bar{z}} = \frac{\partial f}{\partial x}\frac{\partial x}{\partial \bar{z}} + \frac{\partial f}{\partial y}\frac{\partial y}{\partial \bar{z}} = \frac{1}{2}\frac{\partial f}{\partial x} - \frac{1}{2i}\frac{\partial f}{\partial y}.$$

This shows that it is reasonable to define $\partial/\partial z$ and $\partial/\partial \bar{z}$ as we have done. It is then immediately clear that u, v satisfy the Cauchy–Riemann equations if and only if

$$\frac{\partial f}{\partial \bar{z}} = 0.$$

(Carry out in detail.) Thus f is analytic if and only if $\partial f/\partial \bar{z} = 0$.

In Chapter I, §6 we had introduced the **associated vector field**

$$\bar{F}(x, y) = (u(x, y), -v(x, y)).$$

Recall that in calculus courses, one defines a **potential function** for $\bar{F}$ to be a function φ such that

$$\frac{\partial \varphi}{\partial x} = u \quad \text{and} \quad \frac{\partial \varphi}{\partial y} = -v.$$

Theorem 1.2. *Let g be a primitive for f on U, that is, $g' = f$. Write g in terms of its real and imaginary parts,*

$$g = \varphi + i\psi.$$

Then φ is a potential function for $\bar{F}$.

Proof. Go back to Chapter I, §6. By definition, $g' = u + iv$. The first computation of that section shows that

$$\frac{\partial \varphi}{\partial x} = u \quad \text{and} \quad \frac{\partial \varphi}{\partial y} = -v,$$

as desired.

We shall prove shortly that any harmonic function is locally the real part of an analytic function. *In that light, the problem of finding a primitive for an analytic function is equivalent to the problem of finding a potential function for its associated vector field.*

The next theorem gives us the **uniqueness of a harmonic function with prescribed boundary value.**

Theorem 1.3. *Let U be a bounded open set. Let u, v be two continuous functions on the closure U^c of U, and assume that u, v are harmonic on U. Assume that $u = v$ on the boundary of U. Then $u = v$ on U.*

Proof. Subtracting the two harmonic functions having the same boundary value yields a harmonic function with boundary value 0. Let u be such a function. We have to prove that $u = 0$. Suppose there is a point $(x_0, y_0) \in U$ such that $u(x_0, y_0) > 0$. Let

$$\psi(x, y) = u(x, y) + \epsilon x^2 \quad \text{for} \quad (x, y) \in U^c.$$

We use repeatedly the fact that $|x|$ is bounded for $(x, y) \in U^c$. Then

$$\psi(x_0, y_0) > 0$$

for ϵ small enough, and ψ is continuous on U^c, so ψ has a maximum on U^c. For ϵ small, the maximum of ψ is close to the maximum of u itself, and in particular is positive. But $u(x, y) = 0$ for (x, y) on the boundary, and ϵx^2 is small on the boundary for ϵ small. Hence the maximum of ψ must be an interior point (x_1, y_1). It follows that

$$D_1^2\psi(x_1, y_1) \leqq 0 \qquad \text{and} \qquad D_2^2\psi(x_1, y_1) \leqq 0.$$

But

$$(D_1^2 + D_2^2)u = 0 \qquad \text{so} \qquad (D_1^2 + D_2^2)\psi(x_1, y_1) = 2\epsilon > 0.$$

This contradiction proves the uniqueness.

Remarks. In practice, the above uniqueness is weak for two reasons. First, many natural domains are not bounded, and second the function may be continuous on the boundary except at a finite number of points. In examples below, we shall see some physical situations with discontinuities in the temperature function. Hence it is useful to have a more general theorem, which can be obtained as follows.

As to the unboundedness of the domain, it is usually possible to find an isomorphism of a given open set with a bounded open set such that the boundary curves correspond to each other. We shall see an example of this in the Riemann mapping theorem, which gives such isomorphisms with the unit disc. Thus the lack of boundedness of the domain may not be serious.

As to discontinuities on the boundary, let us pick for concreteness the unit disc D. Let u, v be two functions on D^c which are harmonic on the interior D, and which are continuous on the unit circle except at a finite number of points, where they are not defined. Suppose that u and v are equal on the boundary except at those exceptional points. Then the function $u - v$ is harmonic on the open disc D, and is continuous with value 0 on the boundary except at a finite number of points where it is not defined. Thus for the uniqueness, we need the following generalization of Theorem 1.3.

Theorem 1.4. *Let u be a bounded function on the closed unit disc D^c. Assume that u is harmonic on D and continuous on the unit circle except at a finite number of points. Assume that u is equal to 0 on the unit circle except at a finite number of points. Then $u = 0$ on the open disc D.*

The situation is similar to that of removable singularities in Chapter V, §3, but technically slightly more difficult to deal with. We omit the proof. If one does not assume that u is bounded, then the conclusion of the theorem is not true in general. See Exercise 6.

Application: Perpendicularity

Recall from the calculus of several variables (actually two variables) that

$$\operatorname{grad} u = (D_1 u, D_2 u) = \left(\frac{\partial u}{\partial x}, \frac{\partial u}{\partial y}\right).$$

Let c be a number. The equation

$$u(x, y) = c$$

is interpreted as the equation of the level curve, consisting of those points at which u takes the constant value c. If u is interpreted as a potential function, these curves are called curves of **equipotential**. If u is interpreted as temperature, these curves are called **isothermal** curves. Except for such fancy names, they are just level curves of the function u. From calculus, you should know that grad $u(x, y)$ is perpendicular (orthogonal) to the curve at that point, as illustrated on the figure.

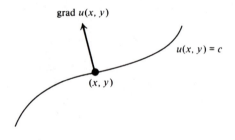

Figure 1

Let $A = (a_1, a_2)$ and $B = (b_1, b_2)$ be vectors. You should know their dot product,

$$A \cdot B = a_1 b_1 + a_2 b_2,$$

and you should know that A is perpendicular to B if and only if their dot product is equal to 0.

Using this and the chain rule, we recall the proof that the gradient is perpendicular to the level curve. We suppose that the level curve is parametrized, i.e. given in the form $\gamma(t)$ for t in some interval. Then we have

$$u(\gamma(t)) = c \qquad \text{for all } t.$$

Differentiating with respect to t yields by the chain rule

$$\operatorname{grad} u(\gamma(t)) \cdot \gamma'(t) = 0,$$

which proves what we wanted.

The following statement is an immediate consequence of the Cauchy–Riemann equations.

Let $f = u + iv$ be analytic. Then grad u *and* grad v *are perpendicular.*

Indeed, we take the dot product of

$$\left(\frac{\partial u}{\partial x}, \frac{\partial u}{\partial y}\right) \quad \text{and} \quad \left(\frac{\partial v}{\partial x}, \frac{\partial v}{\partial y}\right)$$

and apply the Cauchy–Riemann equations to find the value 0.

Two curves $u = c$ and $v = c'$ are said to be **perpendicular** at a point (x, y) if grad u is perpendicular to grad v at (x, y). Hence the above statement is interpreted as saying:

The level curves of the real part and imaginary part of an analytic function are perpendicular (or in other words, intersect orthogonally).

In this case when u is given as the potential function arising from two point sources of electricity, then the level curves for u and v look like Fig. 2.

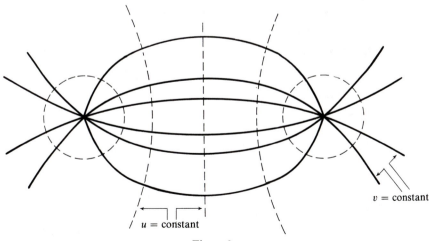

v = constant

u = constant

Figure 2

Application: Flow Lines

For each point (x, y) in the plane, we have an associated vector

$$(x, y) \mapsto \operatorname{grad} u(x, y).$$

This association defines what is called a **vector field**, which we may visualize as arrows shown on Fig. 3.

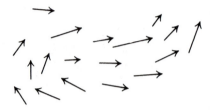

Figure 3

Let us abbreviate

$$G(x, y) = \operatorname{grad} u(x, y).$$

An **integral curve** for the vector field G is a curve γ such that

$$\gamma'(t) = G(\gamma(t)).$$

This means that the tangent vector at every point of the curve is the prescribed vector by G. Such an integral curve is shown on Fig. 4. If we interpret the vector field G as a field of forces, then an integral curve is the path over which a bug will travel, when submitted to such a force.

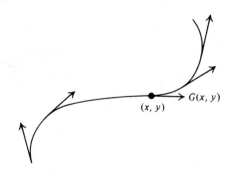

Figure 4

Suppose that $f = u + iv$ is analytic, as usual. We have seen that the level curves of v are orthogonal to the level curves of u. Thus the level curves of v have the same direction as the gradient of u. It can be shown from the uniqueness of the solutions of differential equations that the level curves of v are precisely the integral curves of the vector field $G = \operatorname{grad} u$. Thus interpreting u as temperature, for instance, we may say:

If $u = \operatorname{Re} f$ and $v = \operatorname{Im} f$, where f is analytic, then the heat flow of the temperature function u occurs along the level curves of v.

Finally, let U be simply connected, and let

$$f: U \to H$$

be an isomorphism of U with the upper half plane. We write $f = u + iv$ as usual. The curves

$$v = \text{constant}$$

in H are just horizontal straight lines. The level curves of v in U therefore correspond to these straight lines under the function f.

Consider the example of Chapter VII, §4 given by

$$f(z) = z + 1/z.$$

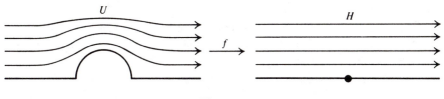

Figure 5

You should have worked out that this gives an isomorphism as shown on the figure. We interpret the right-hand figure as that of a fluid flowing horizontally in the upper half plane, without obstacle. The bump provided by the semicircle in the left-hand figure provides an obstacle to the flow in the open set U.

The nature of the physical world is such that the flow lines on the left are exactly the lines corresponding to the horizontal lines on the right under the mapping function f! Thus the flow lines on the left are exactly the level curves $v = $ constant in U.

This shows how an isomorphism

$$f: U \to H$$

can be applied to finding flow lines. The same principle could be applied to a similar obstacle as in Fig. 6. The open set U is defined here as that portion of the upper half plane obtained by deleting the vertical segment $(0, y)$ with $0 < y < \pi$ from the upper half plane. As an exercise, determine the isomorphism f to find the flow lines in U. Cf. Exercises 7, 8 and the examples of Chapter VII, §4.

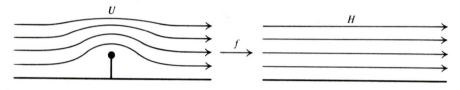

Figure 6

EXERCISES VIII §1

1. Let $\Delta = \left(\dfrac{\partial}{\partial x}\right)^2 + \left(\dfrac{\partial}{\partial y}\right)^2$. Verify that

$$\Delta = 4 \frac{\partial}{\partial z} \frac{\partial}{\partial \bar{z}}.$$

2. Let f be analytic, and $\bar{f} = u - iv$ the complex conjugate function. Verify that $\partial \bar{f}/\partial z = 0$.

3. Let $f: U \to V$ be an analytic isomorphism, and let φ be a harmonic function on V, which is the real part of an analytic function. Prove that the composite function $\varphi \circ f$ is harmonic.

4. Prove that the imaginary part of an analytic function is harmonic.

5. Prove the uniqueness statement in the following context. Let U be an open set contained in a strip $a \leq x \leq b$, where a, b are fixed numbers, and as usual $z = x + iy$. Let u be a continuous function on U^c, harmonic on U. Assume that u is 0 on the boundary of U, and

$$\lim u(x, y) = 0$$

as $y \to \infty$ or $y \to -\infty$, uniformly in x. In other words, given ϵ there exists $C > 0$ such that if $y > C$ or $y < -C$ and $(x, y) \in U$ then $|u(x, y)| < \epsilon$. Then $u = 0$ on U.

6. Let

$$u(x, y) = \operatorname{Re} \frac{i + z}{i - z} \quad \text{for } z \neq i \quad \text{and} \quad u(0, 1) = 0.$$

Show that u harmonic on the unit disc, is 0 on the unit circle, and is continuous on the closed unit disc except at the point $z = i$. This gives a counterexample to the uniqueness when u is not bounded.

7. Find an analytic function whose real part is the given function.

(a) $u(x, y) = 3x^2 y - y^3$ (b) $x - xy$

(c) $\dfrac{y}{x^2 + y^2}$ (d) $\log\sqrt{x^2 + y^2}$

(e) $\dfrac{y}{(x - t)^2 + y^2}$ where t is some real number.

8. Let $f(z) = \log z$. If $z = re^{i\theta}$, then

$$f(z) = \log r + i\theta,$$

so the real parts and imaginary parts are given by

$$u = \log r \qquad \text{and} \qquad v = \theta.$$

Draw the level curves $u = $ constant and $v = $ constant. Observe that they intersect orthogonally.

9. Let V be the open set obtained by deleting the segment $[0, 1]$ from the right half plane, as shown on the figure. In other words V consists of all complex numbers $x + iy$ with $x > 0$, with the exception of the numbers $0 < x \leqq 1$.
   (a)   What is the image of V under the map $z \mapsto z^2$.
   (b)   What is the image of V under the map $z \mapsto z^2 - 1$?
   (c)   Find an isomorphism of V with the right half plane, and then with the upper half plane. [*Hint*: Consider the function $z \mapsto \sqrt{z^2 - 1}$.]

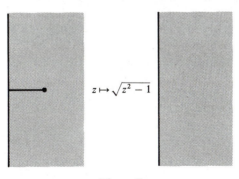

Figure 7

10. Let U be the open set discussed at the end of the section, obtained by deleting the vertical segment of points $(0, y)$ with $0 \leqq y \leqq 1$ from the upper half plane. Find an analytic isomorphism

$$f: U \mapsto H.$$

[*Hint*: Rotate the picture by 90° and use Exercise 9.]

11. Let U be an open set and let $z_0 \in U$. The **Green's function for U originating at** z_0 is a real function g defined on the closure U^c of U, continuous except at z_0, and satisfying the following conditions:

GR 1. $g(z) = \log |z - z_0| + \psi(z)$, where ψ is harmonic on U.

GR 2. g vanishes on the boundary of U.

(a) Prove that a Green's function is uniquely determined if U is bounded.

(b) Let U be simply connected, with smooth boundary. Let

$$f: U \to D$$

be an analytic isomorphism of U with the unit disc such that $f(z_0) = 0$. Let

$$g(z) = \operatorname{Re} \log f(z).$$

Show that g is a Green's function for U. You may assume that f extends to a continuous function from the boundary of U to the boundary of D.

VIII §2. Examples

We shall give examples, some of which are formulated in physical terms. Let U be an open set whose boundary is a smooth curve C. We shall assume throughout that any physical function mentioned is harmonic. In physics, functions like temperature, potential functions, are harmonic. We shall also assume that the uniqueness of a harmonic function on U with prescribed boundary value holds if the boundary value is assumed bounded and continuous except at a finite number of points.

Our examples are constructed for special open sets, which are simply connected. In general any such set is analytically isomorphic to the disc, or preferably to the upper half plane. Let

$$f: U \to H$$

be such an isomorphism. In practice, it is clear how f behaves at the boundary of U, and how it maps this boundary on the boundary of H, i.e. on the real axis. To construct a harmonic function on U with prescribed boundary values, it therefore suffices to construct a function φ on H, and then take the composite $\varphi \circ f$ (see Exercise 3 of the preceding section). In practice, there always exist nice explicit formulas giving the isomorphism f.

Example. We wish to describe the temperature in the upper half plane if the temperature is fixed with value 0 on the positive real axis, and fixed with value 20 on the negative real axis. As mentioned, temperature $v(z)$ is assumed to be harmonic.   We recall that we can define

$$\log z = r(z) + i\theta(z)$$

for z in any simply connected region, in particular for

$$0 \leq \theta(z) \leq \pi, \qquad r(z) > 0$$

omitting the origin.   We have

$$\theta(z) = 0 \text{ if } z \text{ is on the positive real axis,}$$

$$\theta(z) = \pi \text{ if } z \text{ is on the negative real axis.}$$

The function θ (sometimes denoted by arg) is the imaginary part of an analytic function, and hence is harmonic.   The desired temperature is therefore obtainable as an appropriate constant multiple of $\theta(z)$, namely

$$v(z) = \frac{20}{\pi} \theta(z) = \frac{20}{\pi} \arg z.$$

In terms of x, y we can also write

$$\theta(z) = \arctan \ y/x.$$

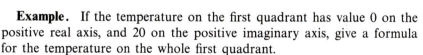

Figure 8

Example. If the temperature on the first quadrant has value 0 on the positive real axis, and 20 on the positive imaginary axis, give a formula for the temperature on the whole first quadrant.

Again we seek a harmonic function having the desired boundary values.   We reduce the problem to the preceding example by using an analytic isomorphism between the first quadrant and the upper half plane, namely

$$z \mapsto z^2.$$

Therefore the solution of the problem in the present instance is given by

$$T(z) = \frac{20}{\pi} \arg z^2 = \frac{40}{\pi} \arctan y/x$$

if $z = x + iy$, and z lies in the first quadrant.

Example. We assume that you have worked Exercise 5. Let A be the upper semidisc. We wish to find a harmonic function φ on A which has value 20 on the positive real axis bounding the semidisc, and value 0 on the negative real axis bounding the semidisc. Furthermore, we ask that $\partial\varphi/\partial n = 0$ on the semicicrcle.

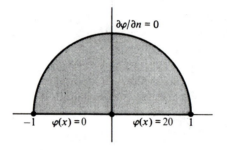

Figure 9

The analytic function $\log z$ maps the semidisc on the horizontal strip as shown on Fig. 10. We may therefore solve the problem with some function v on the strip, and take $\varphi(z) = v(\log z)$ as the solution on the semidisc.

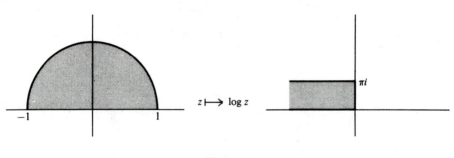

Figure 10

The semicircle is mapped on the vertical segment bounding the strip, and the condition $\partial\varphi/\partial n = 0$ on this segment means that if we view v as a function of two variables (x, y) then $\partial v/\partial x = 0$. Thus v is a function of y alone, and it must have the value 20 on the negative real axis, value 0 at any point $x + \pi i$. Such a function is

$$v(x, y) = 20 - \frac{20}{\pi} y.$$

Consequently,

$$\varphi(z) = 20 - \frac{20}{\pi} v(\log z) = 20 - \frac{20}{\pi} \theta(z),$$

where $0 \leqq \theta(z) \leqq \pi$.

Remark. The condition $\partial \varphi / \partial n = 0$ along a curve is usually interpreted physically as meaning that the curve is insulated, if the harmonic function is interpreted as temperature.

Exercises VIII §2

1. Find a harmonic function on the upper half plane with value 1 on the positive real axis and value -1 on the negative real axis.

2. Find a harmonic function on the indicated region, with the boundary values as shown.

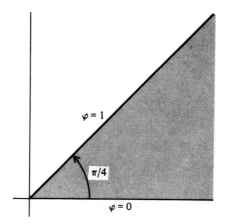

Figure 11

3. Find the temperature on a semicircular plate of radius 1, as shown on the figure, with the boundary values as shown. Value 0 on the semicircle, value 1 on one segment, value 0 on the other segment.

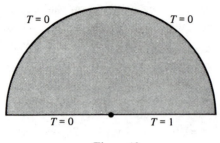

$T = 0$ $T = 0$

$T = 0$ $T = 1$

Figure 12

4. Find a harmonic function on the unit disc which has the boundary value 0 on the lower semicircle and boundary value 1 on the upper semicircle.

In the next exercise, recall that a function $\varphi: U \to \mathbf{R}$ is said to be of class C^1 if its partial derivatives $D_1\varphi$ and $D_2\varphi$ exist and are continuous. Let V be another open set. A mapping

$$f: V \to \mathbf{R}^2$$

where $f(x, y) = (u(x, y), v(x, y))$ is said to be of class C^1 if the two coordinate functions u, v are of class C^1.

If $\eta: [a, b] \to V$ is a curve in V, then we may form the composite curve $f \circ \eta$ such that

$$(f \circ \eta)(t) = f(\eta(t)).$$

Then $\gamma = f \circ \eta$ is a curve in U. Its coordinates are

$$f(\eta(t)) = (\eta(t)), v(\eta(t))).$$

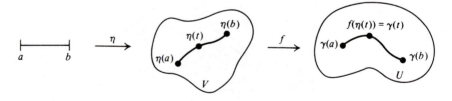

Figure 13

5. Let $\gamma: [a, b] \to \mathbf{R}^2$ be a smooth curve. Let

$$\gamma(t) = (\gamma_1(t), \gamma_2(t))$$

be the expression of γ in terms of its coordinates. The **tangent vector** is given by the derivative $\gamma'(t) = (\gamma_1'(t), \gamma_2'(t))$. We define

$$N(t) = \left(\gamma_2'(t), \, -\gamma_1'(t)\right)$$

to be the **normal vector**. We define the **unit normal vector** to be

$$\mathbf{n}(t) = \frac{N(t)}{|N(t)|}, \qquad \text{where} \quad |N(t)| = \sqrt{N_1(t)^2 + N_2(t)^2},$$

assuming throughout that $|\gamma'(t)| \neq 0$ for all t. Verify that $\gamma'(t) \cdot N(t) = 0$.
If γ is a curve in an open set U, and φ is of class C^1 on U, we define

$$\frac{\partial \varphi}{\partial n} = (\operatorname{grad} \varphi) \cdot \mathbf{n}$$

to be the **outward normal derivative** of φ along the curve.

(a)   Prove that if $\partial \varphi / \partial n = 0$, then this condition remains true under a change of parametrization.

(b)   Let

$$f: V \to U$$

be a C^1-mapping. Let η be a curve in V and let $\gamma = f \circ \eta$. If $\partial \varphi / \partial n = 0$ on γ, show that

$$\partial(\varphi \circ f)/\partial n = 0 \text{ on } \eta.$$

6. Find a harmonic function φ on the indicated regions, with the indicated boundary values. (Recall what $\sin z$ does to a vertical strip.)

(a)

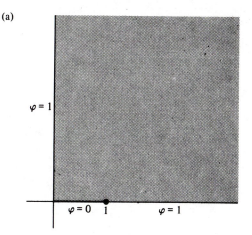

$\varphi = 1$

$\varphi = 0$ 1 $\varphi = 1$

Figure 14

(b)

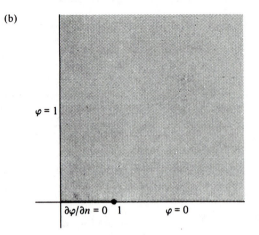

(c)

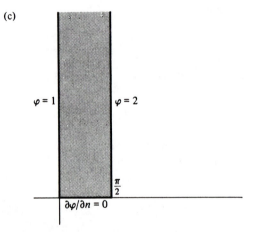

(d)

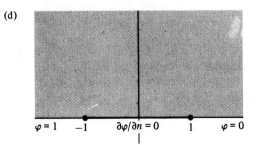

(e)

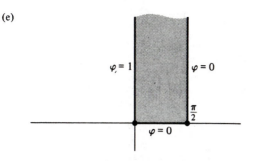

Figure 14 (*continued*)

VIII §3. Basic Properties of Harmonic Functions

In §1 we saw that the real part of an analytic function is harmonic. Here we prove the converse.

> **Theorem 3.1.** *Let U be a connected and simply connected open set. Let u be harmonic on U. Then there exists an analytic function f on U such that $u = \operatorname{Re} f$. The difference of two such functions is a pure imaginary constant.*

Proof. Let

$$h = \frac{\partial u}{\partial x} - i\frac{\partial u}{\partial y}.$$

Then h has continuous partials of first order. Furthermore h is analytic. To see this, it suffices to prove that the real and imaginary parts of h satisfy the Cauchy–Riemann equations, that is

$$\frac{\partial}{\partial x}\left(\frac{\partial u}{\partial x}\right) = -\frac{\partial}{\partial y}\left(\frac{\partial u}{\partial y}\right) \quad \text{and} \quad \frac{\partial}{\partial y}\frac{\partial u}{\partial x} = -\frac{\partial}{\partial x}\left(-\frac{\partial u}{\partial y}\right).$$

The first relation holds by the definition of harmonicity, and the second relation holds because taking partial derivatives with respect to y and with respect to x commutes. Hence h is analytic. Since U is assumed simply connected, h has a primitive f on U, so $f'(z) = h(z)$ for all $z \in U$. Let

$$f = v + iw,$$

so $v = \operatorname{Re} f$ and $w = \operatorname{Im} f$. Then

$$f' = \frac{\partial v}{\partial x} - i\frac{\partial v}{\partial y},$$

so u and v have the same partial derivatives. It follows that there is a constant C such that $u = v + C$. [*Proof:* Let z_0 be any point of U, and let $\gamma: [a, b] \to U$ be a curve joining z_0 with a point z in U. Let $g = u - v$. Then the partial derivatives of g are 0. By the chain rule we have

$$\frac{d}{dt} g(\gamma(t)) = \frac{\partial g}{\partial x} \frac{\partial x}{\partial t} + \frac{\partial g}{\partial y} \frac{\partial y}{\partial t} = 0.$$

Hence $g(\gamma(t))$ is constant, so $g(z_0) = g(z)$. This is true for all points $z \in U$, whence g is constant, as desired.] Subtracting the constant C from f yields the desired analytic function function having the given real part u, and proves the existence.

As to uniqueness, let f, g be analytic functions having the same real part. Then the real part of $f - g$ is 0. It follows that $f - g$ is locally constant (for instance because an analytic function is an open mapping, so its image cannot lie on a straight line). Hence f is constant because U is connected, Theorem 1.2 of Chapter III, and this constant is pure imaginary since its real part is 0.

Theorem 3.2 (Mean Value Theorem). *Let u be a harmonic function on an open set U. Let $z_0 \in U$, and let $r > 0$ be a number such that the closed disc of radius r centered at z_0 is contained in U. Then*

$$u(z_0) = \frac{1}{2\pi} \int_0^{2\pi} u(z_0 + re^{i\theta}) \, d\theta.$$

Proof. There is a number $r_1 > r$ such that the disc of radius r_1 centered at z_0 is contained in U. Any $r_1 > r$ and close to r will do. By Theorem 3.1, there is an analytic function f on the disc of radius r_1 such that $u = \operatorname{Re} f$. By Cauchy's theorem,

$$f(z_0) = \frac{1}{2\pi i} \int_C \frac{f(\zeta)}{\zeta - z_0} \, d\zeta,$$

where C is the circle of radius r centered at z_0.

We parametrize the circle by $\zeta = z_0 + re^{i\theta}$, so $d\zeta = ire^{i\theta} \, d\theta$. The integral then gives

$$f(z_0) = \frac{1}{2\pi} \int_0^{2\pi} f(z_0 + re^{i\theta}) \, d\theta.$$

If we write $f = u + iv$, then the desired relation falls out for the real part u. This concludes the proof.

Theorem 3.3. *Let u be harmonic on a connected open set U. Suppose that u has a maximum at a point z_0 in U. Then u is constant.*

Proof. First we prove that u is constant in a disc centered at z_0. By Theorem 3.1 there is an analytic function f on such a disc such that Re $f = u$. Then $e^{f(z)}$ is analytic, and

$$|e^{f(z)}| = e^{u(z)}.$$

Since the exponential function is strictly increasing, it follows that a maximum for u is also a maximum for e^u, and hence also a maximum of $|e^f|$. By the maximum modulus principle for analytic functions, it follows that e^f is constant on the disc. Then e^u is constant, and finally u is constant, thus proving the theorem locally.

We now extend the theorem to an arbitrary connected open set. Let S be the set of points z in U such that u is constant in a neighborhood of z. Then S contains z_0, and S is open. By Theorem 1.6 of Chapter III, it will suffice to prove that S is closed in U. So let z_1 be a point in the closure of S, and z_1 contained in U. Let V be an open disc centered at z_1 and contained in U. By Theorem 3.1 there is an analytic function f on V such that Re $f = u$. Also, since z_1 is in the closure of S, there is a point $z_2 \in S$ and also $z_2 \in V$. By the definition of S, u is constant in some neighborhood of z_2, or in other words, Re f is constant in some neighborhood of z_2. By the open mapping theorem, this implies that f itself is constant in some neighborhood of z_2. Since V is connected, it follows that f is constant on V, so u is constant on V. This concludes concludes the proof.

EXERCISE VIII §3

1. **Green's theorem** in calculus states: *Let $p = p(x, y)$ and $q = q(x, y)$ be C^1 functions on the closure of an open set U which is the inside of a simple closed curve C, oriented counterclockwise. Then*

$$\int_C p\,dx + q\,dy = \iint_A \left(\frac{\partial q}{\partial x} - \frac{\partial p}{\partial y}\right) dy\,dx,$$

where A is the inside of the curve.

Suppose that f is analytic on A and on its boundary. Show that Green's theorem implies Cauchy's theorem, assuming that the real and imaginary parts of f are of class C^1, i.e. show that

$$\int_C f = 0.$$

VIII §4. Construction of Harmonic Functions

Let U be a simply connected open set with smooth boundary. By the Riemann mapping theorem, there is an analytic isomorphism of U with the unit disc, extending to a continuous isomorphism at the boundary. To construct a harmonic function on U with prescribed boundary value, it suffices therefore to do so for the disc.

In this case, we use the method of Dirac sequences, or rather Dirac families. We recall what that means. We shall deal with periodic functions of period 2π in the sequel, so we make that assumption from the beginning. By a **Dirac sequence** we shall mean a sequence of functions $\{K_n\}$ of a real variable, periodic of period 2π, real valued, satisfying the following properties.

DIR 1. *We have $K_n(x) \geq 0$ for all n and all x.*

DIR 2. *Each K_n is continuous, and*

$$\int_0^{2\pi} K_n(t)\, dt = 1.$$

DIR 3. *Given ϵ and δ, there exists N such that if $n \geq N$, then*

$$\int_{-\pi}^{-\delta} K_n + \int_\delta^\pi K_n < \epsilon.$$

Condition **DIR 2** means that the area under the curve $y = K_n(x)$ is equal to 1. Condition **DIR 3** means that this area is concentrated near 0 if n is taken sufficiently large. Thus a family $\{K_n\}$ as above looks like Fig. 15. The functions $\{K_n\}$ have a peak near 0. In the applications, it

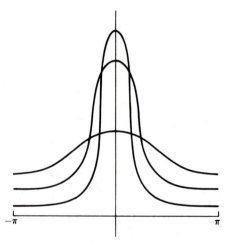

Figure 15

is also true that the functions K_n are even, that is, $K_n(-x) = K_n(x)$, but we won't need this. If f is any periodic function, we define the **convolution** with K_n to be

$$f_n(x) = K_n * f(x) = \int_{-\pi}^{\pi} f(t)K_n(x - t)\, dt.$$

Theorem 4.1. *Let f be continuous periodic. Then the sequence $\{K_n * f\}$ converges to f uniformly.*

Proof. Changing variables, we have

$$f_n(x) = \int_{-\pi}^{\pi} f(x - t)K_n(t)\, dt.$$

On the other hand, by **DIR 2**,

$$f(x) = f(x) \int_{-\pi}^{\pi} K_n(t)\, dt = \int_{-\pi}^{\pi} f(x)K_n(t)\, dt.$$

Hence

$$f_n(x) - f(x) = \int_{-\pi}^{\pi} [f(x - t) - f(x)]K_n(t)\, dt.$$

By the compactness of the circle, and the uniform continuity of f, we conclude that given ϵ there is δ such that whenever $|t| < \delta$ we have

$$|f(x - t) - f(x)| < \epsilon$$

for all x. Let B be a bound for f. Then we select N such that if $n \geq N$,

$$\int_{-\pi}^{-\delta} + \int_{\delta}^{\pi} K_n < \frac{\epsilon}{2B}.$$

We have

$$|f_n(x) - f(x)| \leq \int_{-\pi}^{-\delta} + \int_{-\delta}^{\delta} + \int_{\delta}^{\pi} |f(x - t) - f(x)|K_n(t)\, dt.$$

To estimate the first and third integral, we use the given bound B for f so that $|f(x - t) - f(x)| \leq 2B$. We obtain

$$\int_{-\pi}^{-\delta} + \int_{\delta}^{\pi} |f(x - t) - f(x)|K_n(t)\, dt \leq 2B \int_{-\pi}^{-\delta} + \int_{\delta}^{\pi} K_n(t)\, dt < \epsilon.$$

For the integral in the middle, we have the estimate

$$\int_{-\delta}^{\delta} |f(x-t) - f(x)| K_n(t)\, dt \leq \int_{-\delta}^{\delta} \epsilon K_n \leq \int_{-\pi}^{\pi} \epsilon K_n \leq \epsilon.$$

This proves our theorem.

We leave it as an exercise to prove that if K_n is of class C^1,

$$\frac{d}{dx}(K_n * f)(x) = \left(\frac{dK_n}{dx}\right) * f.$$

This is merely differentiating under the integral sign.

We shall work with polar coordinates r, θ. It is an exercise to see that the Laplace operator can be put in polar coordinates by

$$\boxed{\Delta = \left(\frac{\partial}{\partial x}\right)^2 + \left(\frac{\partial}{\partial y}\right)^2 = \frac{\partial^2}{\partial r^2} + \frac{1}{r}\frac{\partial}{\partial r} + \frac{1}{r^2}\frac{\partial^2}{\partial \theta^2}.}$$

It was convenient to formulate the general Dirac property for sequences, but we shall work here with families, indexed by r with $0 < r < 1$ and r tending to 1, rather than n tending to infinity. We define the **Poisson kernel** as

$$P_r(\theta) = \frac{1}{2\pi} \sum_{k=-\infty}^{\infty} r^{|k|} e^{ik\theta}.$$

The series is absolutely convergent, and uniformly so, dominated by a geometric series, if r stays away from 1. Simple trigonometric identities show that

$$(*) \qquad\qquad P_r(\theta) = \frac{1}{2\pi}\frac{1-r^2}{1-2r\cos\theta + r^2}.$$

The smallest value of the denominator occurs when $\cos\theta = 1$, and we therefore see that

DIR 1. $P_r(\theta) \geq 0$ for all r, θ.

Integrating the series term by term yields

DIR 2. $$\int_{-\pi}^{\pi} P_r(\theta)\, d\theta = 1.$$

Finally, we leave it to the reader as a simple estimate using expression (∗) for $P_r(\theta)$ to prove:

DIR 3. *Given ϵ and δ there exists r_0, $0 < r_0 < 1$, such that if*

$$r_0 < r < 1,$$

then

$$\int_{-\pi}^{-\delta} P_r + \int_{\delta}^{\pi} P_r < \epsilon.$$

Thus we view $\{P_r\}$ as a Dirac family, with $r \to 1$. We let

$$f_r = P_r * f$$

and write $f(r, \theta) = f_r(\theta)$. Then $f_r(\theta)$ is a function on the *open* disc. Theorem 4.1 yields:

$$f_r(\theta) \to f(\theta) \qquad \text{uniformly as} \quad r \to 1.$$

Theorem 4.2. *The function $(r, \theta) \mapsto f_r(\theta)$ is harmonic on the disc.*

Proof. The Laplace operator in polar coordinates can be applied to $P_r(\theta)$, differentiating the series term by term, which is obviously allowable. If you do this, you will find that

$$\left[r^2 \left(\frac{\partial^2}{\partial r^2} \right) + r \left(\frac{\partial}{\partial r} \right) + \frac{\partial^2}{\partial \theta^2} \right] P_r(\theta) = 0.$$

Thus $\Delta P = 0$, where P denotes the function of two variables,

$$P(r, \theta) = P_r(\theta).$$

Differentiating under the integral sign, we then obtain

$$\Delta((P_r * f)(\theta)) = (\Delta P_r(\theta)) * f = 0.$$

This proves the theorem.

We view the original periodic function f as a boundary value on the circle. The function

$$u(r, \theta) = f_r(\theta) = P_r * f(\theta)$$

is defined by convolution for $0 \leq r < 1$, and by continuity for $r = 1$. The theorem yields the existence of a harmonic function u having the prescribed value f on the circle.

In many applications, e.g. physical applications, but even theoretical ones, it is not convenient to assume that the boundary value function is continuous. One should allow for at least a finite number of discontinuities, although still assuming that the function is bounded. In that case, an analysis of the proof shows that as much as one would expect of the theorem remains true, i.e. the reader will verify that the proof yields:

Theorem 4.3. *Let f be a bounded function on the reals, periodic of period 2π. Let S be a compact set where f is continuous. Then the sequence $\{K_n * f\}$ converges uniformly to f on S. The function*

$$u(r, \theta) = f_r(\theta) = P_r * f(\theta)$$

is harmonic on the open disc $0 \leq r < 1$ and $0 \leq \theta \leq 2\pi$.

EXERCISES VIII §4

One can also consider Dirac sequences or families over the whole real line. We use a notation which will fit a specific application. For each $y > 0$ suppose given a continuous function P_y on the real line, satisfying the following conditions:

DIR 1. *$P_y(t) \geq 0$ for all y, and all real t.*

DIR 2.
$$\int_{-\infty}^{\infty} P_y(t)\, dt = 1.$$

DIR 3. *Given ϵ, δ there exists $y_0 > 0$ such that if $0 < y < y_0$, then*

$$\int_{-\infty}^{-\delta} + \int_{\delta}^{\infty} P_y(t)\, dt < \epsilon.$$

We call $\{P_y\}$ a **Dirac family** again, for $y \to 0$. Prove:

1. Let f be continuous on **R**, and bounded. Define the convolution $P_y * f$ by

$$P_y * f(x) = \int_{-\infty}^{\infty} P_y(x - t) f(t)\, dt.$$

Prove that $P_y * f(x)$ converges to $f(x)$ as $y \to 0$ for each x where f is continuous.
    The proof should also apply to the case when f is bounded, and continuous except at a finite number of points, etc.

2. Let

$$P_y(t) = \frac{1}{\pi} \frac{y}{t^2 + y^2} \qquad \text{for} \quad y > 0.$$

Prove that $\{P_y\}$ is a Dirac family. It has no special name, like the Poisson family as discussed in the text, but it is classical.

3. Define for all real x and $y > 0$:

$$F(x, y) = P_y * f(x).$$

Prove that F is harmonic. In fact, show that the Laplace operator

$$\left(\frac{\partial}{\partial x}\right)^2 + \left(\frac{\partial}{\partial y}\right)^2$$

applied to

$$\frac{y}{(t - x)^2 + y^2}$$

yields 0.

You will have to differentiate under an integral sign, with the integral being taken over the real line. You can handle this in two ways.

(i)   Work formally and assume everything is OK.

(ii) Justify all the steps. In this case, you have to use a lemma like that proved in Chapter XII, §1.

The above procedure shows how to construct a harmonic function on the upper half plane, with given boundary value, just as was done for the disc in the text.

VIII §5. The Poisson Representation

Theorem 5.1. *Let u be continuous on the closed unit disc D^c, and harmonic on the disc D. Then there exists an analytic function f on D such that $u = \operatorname{Re} f$. In fact*

$$f(z) = \frac{1}{2\pi i} \int_C \frac{\zeta + z}{\zeta - z} u(\zeta) \frac{d\zeta}{\zeta}$$

where C is the unit circle.

Proof. The function f defined by the above integral is obviously analytic on D (differentiate under the integral sign). We have to identify its real part with u. But the integrand is merely another expression for the convolution of the Poisson kernel with u. Indeed, the reader will easily verify that if $z = re^{i\theta}$ is the polar expression for z, then

$$P_r(\theta - t) = \frac{1}{2\pi} \operatorname{Re} \frac{e^{it} + z}{e^{it} - z} = \frac{1}{2\pi} \frac{1 - r^2}{1 - 2r\cos(\theta - t) + r^2}.$$

The unit circle is parametrized by $\zeta = e^{it}$, $d\zeta = e^{it}idt$, and so the expression for the real part of f can also be written

$$\operatorname{Re} f = \frac{1}{2\pi} \int_{-\pi}^{\pi} \operatorname{Re} \frac{e^{it} + z}{e^{it} - z} u(e^{it}) \, dt,$$

which is the convolution integral. Applying Theorem 4.3 shows that the real part $\operatorname{Re} f$ on D extends continuously to the boundary and that its boundary value is precisely u. The uniqueness of harmonic functions with given boundary value shows that the real part of f is also equal to u on the interior, as was to be shown.

The integral expression of Theorem 4.1 gives a bound for f in terms of its real part and $f(0)$. Such a bound can be obtained in a simpler manner just using the maximum modulus principle, and we shall give this other proof in §5 of the next chapter.

EXERCISES VIII §5

1. Extend the results of §4 to discs of arbitrary radius R by means of the Poisson kernel for radius R, namely

$$P_r(\theta) = \frac{1}{2\pi} \frac{R^2 - r^2}{R^2 - 2Rr \cos \theta + r^2}$$

for $0 \leqq r < R$. So prove the statements about a Dirac family, give the existence and uniqueness of a harmonic function with given continuous boundary value on the closed disc of radius R.

2. Prove the inequalities

$$\frac{R - r}{R + r} \leqq 2\pi P_r(\theta - t) \leqq \frac{R + r}{R - r}$$

for $0 \leqq r < R$.

3. For a harmonic function $u \geqq 0$ on the disc $D(\alpha, R)$ with continuous extension to the closed disc $D^c(\alpha, R)$, show that

$$\frac{R - r}{R + r} u(\alpha) \leqq u(\alpha + re^{i\theta}) \leqq \frac{R + r}{R - r} u(\alpha).$$

4. Let $\{u_n\}$ be a sequence of harmonic functions on the open disc. If it converges uniformly on compact subsets of the disc, then the limit is harmonic.

5. Let u be a continuous function on the closure of the upper half plane (i.e. on the upper half plane and on the real line). Assume also that u is harmonic on the

upper half plane, and bounded on the real line. Also assume that u is real valued. Using the Dirac family in the exercises of the preceding section, and the fact that for any analytic function f on the upper half plane, we have the integral formula

$$f(z) = \frac{1}{2\pi i} \int_{-\infty}^{\infty} \frac{f(t)}{t - z} dt,$$

show how to construct an analytic function f on the upper half plane whose real part is u.

If you don't know the integral formula yet (it appeared as Exercise 23, Chapter VI, §1) then prove it too.

PART TWO

VARIOUS ANALYTIC TOPICS

CHAPTER IX

Applications of the Maximum Modulus Principle

We return to the maximum principle in a systematic way, and give several ways to apply it, in various contexts.

One of the most striking applications omitted from standard courses, is to the problem of transcendence: Given some analytic function, describe those points z such that $f(z)$ is an integer, or a rational number, or an algebraic number. This type of question first arose at the end of the nineteenth century, among analysts in Weierstrass' school (Stäckel, Strauss). The question was again raised by Pólya, and Gelfond recognized the connection with the transcendence problems. A bibliography can be found in the books on transcendence by, for instance, Baker, Gelfond, Lang, Schneider. In order not to assume too much for the reader, we shall limit ourselves in this chapter to the case when the function takes on certain values in the rational numbers. The reader who knows about algebraic numbers will immediately see how to extend the proof to that case.

IX §1. The Effect of Zeros, Jensen–Schwarz Lemma

We study systematically the situation when an analytic function has zeros in a certain region. This has the effect of making the function small throughout the region. We want quantitative results showing how small in terms of the number of zeros, possibly counted with their multiplicities. If the function has zeros at points $z_1, \ldots, z_n$ with multiplicities $k_1, \ldots, k_n$, then

$$\frac{f(z)}{(z - z_1)^{k_1} \cdots (z - z_n)^{k_n}}$$

is again analytic, and the maximum modulus principle can be applied to it. On a disc of radius R centered at the origin such that the disc contains all the points $z_1, \ldots, z_n$, suppose that R is large with respect to $|z_1|, \ldots, |z_n|$, say $|z_j| \leq R/2$. Then for $|z| = R$ we get

$$|z - z_j| \geq R/2.$$

If we estimate the above quotient on the circle of radius R (which gives an estimate for this quotient inside by the maximum modulus principle) we obtain an estimate of the form

$$2^{k_1 + \cdots + k_n} \|f\|_R / R^{k_1 + \cdots + k_n}.$$

Specific situations then compare $\|f\|_R$ with the power of R in the denominator to give whatever result is sought. In the applications most often the multiplicities are equal to the same integer. We state this formally.

Theorem 1.1. *Let f be holomorphic on the closed disc of radius R. Let $z_1, \ldots, z_N$ be points inside the disc where f has zeros of multiplicities $\geq M$, and assume that these points lie in the disc of radius R_1. Assume*

$$R_1 \leq R/2.$$

Let $R_1 \leq R_2 \leq R$. Then on the circle of radius R_2 we have the estimate

$$\|f\|_{R_2} \leq \frac{\|f\|_R 4^{MN}}{e^{MN \, \log(R/R_2)}}.$$

Proof. Let $|w| = R_2$. We estimate the function

$$\frac{f(z)}{[(z - z_1) \cdots (z - z_N)]^M} \, [(w - z_1) \cdots (w - z_n)]^M$$

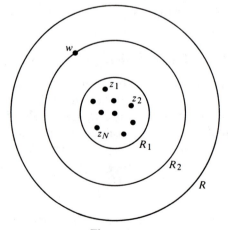

Figure 1

on the circle of radius R. This function has precisely the value $f(w)$ at $z = w$. The estimate $|w - z_j| \leq 2R_2$ is trivial, and the theorem follows at once.

A precise version of Theorem 1.1 will now be given, by using a refinement of the technique of proof.

Theorem 1.2. *Let f be holomorphic on the closed disc of radius R, and assume that $f(0) \neq 0$. Let the zeros of f in the open disc be ordered by increasing absolute value,*

$$z_1, z_2, \ldots, z_N,$$

each zero being repeated according to its multiplicity. Then

$$|f(0)| \leq \frac{\|f\|_R}{R^N} |z_1 z_2 \cdots z_N|.$$

Proof. The function

$$g(z) = f(z) \prod_{n=1}^{N} \frac{R^2 - z\bar{z}_n}{R(z - z_n)}$$

is holomorphic on the closed disc of radius R, and

$$|g(z)| = |f(z)| \qquad \text{when} \quad |z| = R.$$

Hence the maximum modulus principle implies that

$$|g(z)| \leq \|f\|_R,$$

and the theorem follows by putting $z = 0$.

The theorem is known as **Jensen's inequality.** Let $v(r)$ be the number of zeros of f in the closed disc $|z| \leq r$. Then

$$\log \frac{R^N}{|z_1 z_2 \cdots z_N|} = \sum_{n=1}^{N} \int_{|z_n|}^{R} \frac{dx}{x} = \int_{0}^{R} \frac{v(x)}{x} \, dx.$$

Hence we obtain another formulation of **Jensen's inequality,**

$$\boxed{\int_{0}^{R} \frac{v(x)}{x} \, dx \leq \log \|f\|_R - \log |f(0)|.}$$

In many applications, the estimate given by Jensen's inequality suffices, but it is of interest to get the exact relation which exists between the zeros of f and the mean value on the circle.

Theorem 1.3   (Jensen's Formula).   *Let f be analytic on the disc $D = D(0, R)$ of radius R, and assume $f(0) \neq 0$. Let $z_1, z_2, \ldots$ be the zeros of f on D, ordered by increasing absolute value, and repeated according to their multiplicities. Let $r_n = |z_n|$. Then for $r_n \leq r \leq r_{n+1}$ we have*

J 1. $$\log \frac{r^n |f(0)|}{r_1 r_2 \cdots r_n} = \frac{1}{2\pi} \int_0^{2\pi} \log |f(re^{i\theta})| \, d\theta,$$

or also

J 2. $$\log |f(0)| + \int_0^r \frac{v(x)}{x} \, dx = \frac{1}{2\pi} \int_0^{2\pi} \log |f(re^{i\theta})| \, d\theta$$

where $v(x) =$ number of zeros z of f with $|z| \leq x$.

Proof. The second formula follows at once from the first, because we have already seen that

$$\log \frac{r^n}{r_1 r_2 \cdots r_n} = \int_0^R \frac{v(x)}{x} \, dx.$$

We prove the first formula.

Suppose first that f has no zeros in the closed disc $|z| \leq r$. Then $\log f(z)$ is analytic on this disc, and

$$\log f(0) = \frac{1}{2\pi i} \int_{|z|=r} \frac{\log f(z)}{z} \, dz = \frac{1}{2\pi} \int_0^{2\pi} \log f(re^{i\theta}) \, d\theta.$$

Taking the real part proves the theorem in this case.

To deal with the general case, we divide the given function by a polynomial to cancel the zeros, and prove the theorem for each linear factor. We then make use of the obvious fact that if the theorem is true for two functions f, g, then it is valid for their product fg. The details are as follows.

Let

$$g(z) = f(z) \frac{z_1 \cdots z_n}{(z_1 - z) \cdots (z_n - z)},$$

where

$$|z_n| \leq r < |z_{n+1}|.$$

Then $g(0) = f(0)$, and g has no zero on the closed disc of radius r. We have already shown that the theorem is true for g. We can write f as a product of g and linear factors $z_j - z$, as well as constants z_j^{-1}. It will therefore suffice to prove the theorem for the special functions

$$f(z) = z - \alpha, \quad \text{where} \quad \alpha = ae^{i\varphi},$$

with $0 < a \leqq r$. In this case, $|f(0)| = |\alpha| = a$, and the formula to be proved is

$$\log r = \frac{1}{2\pi} \int_0^{2\pi} \log|re^{i\theta} - ae^{i\varphi}|\, d\theta.$$

It will therefore suffice to prove:

Lemma. *If* $0 < a \leqq r$, *then* $\int_0^2 \log|e^{i\theta} - \frac{a}{r}e^{i\varphi}|\, d\theta = 0$.

Proof. Suppose first $a < r$. Then the function

$$\frac{\log\left(1 - \dfrac{a}{r}z\right)}{z}$$

is analytic for $|z| \leq 1$, and from this it is immediate that the desired integral is 0. Next suppose $a = r$, so we have to prove that

(∗) $$\int_0^{2\pi} \log|1 - e^{i\theta}|\, d\theta = 0.$$

To see this we note that the left-hand side is the real part of the complex integral

$$\int_C \log(1 - z)\frac{dz}{iz},$$

taken over the circle of radius 1, $z = e^{i\theta}$, $dz = ie^{i\theta}$. Let $\gamma(\epsilon)$ be the contour as shown on Fig. 2.

For Re $z < 1$ the function

$$\frac{\log(1 - z)}{z}$$

is holomorphic, and so the integral is equal to 0:

$$\int_{\gamma(\epsilon)} \frac{\log(1 - z)}{z}\, dz = 0.$$

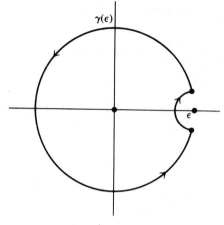

Figure 2

The integral over the small indented circular arc of radius ϵ is bounded in absolute value by a constant times $\epsilon \log \epsilon$, which tends to 0 as ϵ tends to 0. This proves $(*)$, settling the present case.

EXERCISE IX §1

1. Prove Jensen's formula if the function f is also allowed to have poles. In other words, suppose $a_1, \ldots, a_n$ are the zeros and $b_1, \ldots, b_n$ are the poles of f in the disc $|z| \leq r$. Prove that

$$\log r^{m-n} \left| \frac{b_1 \cdots b_n}{a_1 \cdots a_m} f(0) \right| = \frac{1}{2\pi} \int_0^{2\pi} \log |f(re^{i\theta})| \, d\theta.$$

IX §2. The Effect of Small Derivatives

If a function does not have zeros at certain points, but has small derivatives, then it is still true that the function is small in a region not too far away from these points. A quantitative estimate can be given, with a main term which is the same as if the function had zeros, and an error term, measured in terms of the derivatives.

Let $z_1, \ldots, z_N$ be distinct points in the open disc of radius R, and let

$$Q(z) = [(z - z_1) \cdots (z - z_N)]^M.$$

Let f be holomorphic on the closed disc of radius R. Let Γ be the circle of radius R, and let Γ_j be a circle around z_j not containing z_k for $k \neq j$, and contained in the interior of Γ. Then we have for z not equal to any z_j:

Hermite Interpolation Formula

$$\frac{f(z)}{Q(z)} = \frac{1}{2\pi i} \int_{\Gamma} \frac{f(\zeta)}{Q(\zeta)} \frac{d\zeta}{(\zeta - z)}$$

$$- \frac{1}{2\pi i} \sum_{j=1}^{N} \sum_{m=0}^{M-1} \frac{D^m f(z_j)}{m!} \int_{\Gamma_j} \frac{(\zeta - z_j)^m}{Q(\zeta)} \frac{d\zeta}{(\zeta - z)}.$$

This formula, due to Hermite, is a direct consequence of the residue formula. We consider the integral

$$\int_{\Gamma} \frac{f(\zeta)}{Q(\zeta)} \frac{1}{(\zeta - z)} d\zeta.$$

The function has a simple pole at $\zeta = z$ with residue $f(z)/Q(z)$. This gives the contribution on the left-hand side of the formula. The integral is also equal to the sum of integrals taken over small circles around the points $z_1, \ldots, z_N, z$. To find the residue at z_j, we expand $f(\zeta)$ at z_j, say

$$f(\zeta) = a_0 + a_1(\zeta - z_j) + \cdots + a_M(\zeta - z_j)^M + \cdots.$$

Looking at the quotient by $Q(\zeta)$ immediately determines the residues at z_j in terms of coefficients of the expression, which are such that

$$a_m = D^m f(z_j)/m!.$$

The formula then drops out.

It is easy to estimate $|f(z)|$. Multiplying by $Q(z)$ introduces the quotients

$$\frac{Q(z)}{Q(\zeta)} = \prod_{j=1}^{N} \left[\frac{(z - z_j)}{(\zeta - z_j)} \right]^M,$$

which are trivially estimated. The denominator is small according as the radius of Γ_j is small. In applications, one tries to take the Γ_j of not too small radius, and this depends on the minimum distance between the points $z, z_1, \ldots, z_N$. It is a priori clear that if the points are close together, then the information that the function has small derivatives at these points is to a large extent redundant. This information is stronger the wider apart the points are.

Making these estimates, the following result drops out.

Theorem 2.1. *Let f be holomorphic on the closed disc of radius R. Let $z_1, \ldots, z_N$ be distinct points in the disc of radius R_1. Assume*

$$2R_1 < R_2 \quad and \quad 2R_2 < R.$$

Let σ be the minimum of 1, *and the distance between any pair of distinct points among* $z_1, \ldots, z_N$. *Then*

$$\|f\|_{R_2} \leq \frac{\|f\|_R C^{MN}}{(R/R_2)^{MN}} + (CR_2/\sigma)^{MN} \max_{m, j} \frac{|D^m f(z_j)|}{m!},$$

where C is an absolute constant.

An estimate for the derivative of f can then be obtained from Cauchy's formula

$$\frac{D^k f(z)}{k!} = \frac{1}{2\pi i} \int_{|\zeta| = R_2} \frac{f(\zeta)}{(\zeta - z)^{k+1}} \, d\zeta.$$

from which we see that such a derivative is estimated by a similar expression, multiplied by

$$2^k/R_2^k.$$

We may summarize the estimate of Theorem 2.1 by saying that the first term is exactly the same as would arise if f had zeros at the points $z_1, \ldots, z_N$, and the second term is a correcting factor describing the extent to which those points differ from actual zeros. In practice, the derivatives of f are very small at these points, which thus do not differ too much from zeros.

IX §3. Entire Functions with Rational Values

We shall give a theorem showing how the set of points where an entire function takes on rational values is limited. The method follows the classical method of Gelfond and Schneider in proving the transcendence of α^β (when α, β are algebraic $\neq 0$, 1 and β is irrational). It develops the ideas of Schneider, who had partially axiomatized the situation (cf. his book on transcendental numbers), but in a manner which makes the theorem also applicable to a wider class of functions, e.g. abelian functions (cf. Lang's book on transcendental numbers).

We recall that an analytic function is said to be entire if it is analytic on all of **C**. Let $\rho > 0$. We shall say that f has **strict order** $\leq \rho$ if there exists a number $C > 1$ such that for all sufficiently large R we have

$$|f(z)| \leq C^{R^\rho} \qquad \text{whenever} \qquad |z| \leq R.$$

Two functions f, g are called **algebraically independent** if for any polynomial function

$$\sum a_{ij} f^i g^j = 0$$

with complex coefficients a_{ij}, we must have $a_{ij} = 0$ for all i, j.

We denote the rational numbers by $\mathbf{Q}$, as usual. If f, g take on rational values at certain points, we shall construct an auxiliary function $\sum b_{ij} f^i g^j$ which has zeros of high order at these points, and then estimate this latter function.

Theorem 3.1. *Let $f_1,\ldots,f_n$ be entire functions of strict order $\leq \rho$. Assume that at least two of these functions are algebraically independent. Assume that the derivative $d/dz = D$ maps the ring $\mathbf{Q}[f_1,\ldots,f_n]$ into itself, i.e. for each j there is a polynomial P_j with rational coefficients such that*

$$Df_j = P_j(f_1,\ldots,f_n).$$

Let $w_1,\ldots,w_N$ be distinct complex numbers such that

$$f_j(w_i) \in \mathbf{Q}$$

for $j = 1,\ldots,n$ and $i = 1,\ldots,N$. Then $N \leq 4\rho$.

The most classical application of the theorem is to the ring of functions

$$\mathbf{Q}[z, e^z]$$

which is certainly mapped into itself by the derivative $d/dz = D$. The theorem then implies that e^w cannot be rational for any integer $w \neq 0$. For otherwise,

$$e^w, e^{2w}, \ldots, e^{Nw}$$

would be rational, so we would obtain infinitely many numbers at which the exponential takes on rational values, as well as the function z, which contradicts the theorem. The same argument works in the more general case when dealing with algebraic numbers, to show that e^w is not algebraic when w is algebraic $\neq 0$.

Before giving the main arguments proving the theorem, we state some lemmas. The first, due to Siegel, has to do with integral solutions of linear homogeneous equations.

Lemma 1. *Let*

$$a_{11}x_1 + \cdots + a_{1n}x_n = 0,$$

$$\cdots$$

$$a_{r1}x_1 + \cdots + a_{rn}x_n = 0$$

be a system of linear equations with integer coefficients a_{ij}, and $n > r$. Let A be a number such that $|a_{ij}| \leq A$ for all i, j. Then there exists an integral, non-trivial solution with

$$|x_j| \leq 2(2nA)^{r/(n-r)}.$$

Proof. We view our system of linear equations as a linear equation $L(X) = 0$, where L is a linear map, $L: \mathbf{Z}^{(n)} \to \mathbf{Z}^{(r)}$, determined by the matrix of coefficients. If B is a positive number, we denote by $\mathbf{Z}^{(n)}(B)$ the set of vectors X in $\mathbf{Z}^{(n)}$ such that $|X| \leq B$ (where $|X|$ is the maximum of the absolute values of the coefficients of X). Then L maps $\mathbf{Z}^{(n)}(B)$ into $\mathbf{Z}^{(r)}(nBA)$. The number of elements in $\mathbf{Z}^{(n)}(B)$ is $\geq B^n$ and $\leq (2B + 1)^n$. We seek a value of B such that there will be two distinct elements X, Y in $\mathbf{Z}^{(n)}(B)$ having the same image, $L(X) = L(Y)$. For this, it will suffice that $B^n > (2nBA)^r$, and thus it will suffice that $B = (2nA)^{r/(n-r)}$. We take $X - Y$ as the solution of our problem.

The next lemma has to do with estimates of derivatives. By the **size** of a polynomial with rational coefficients we shall mean the maximum of the absolute values of the coefficients. A **denominator** for a set of rational numbers will be any positive integer whose product with every element of the set is an integer. We define in a similar way a denominator for a polynomial with rational coefficients. We abbreviate denominator by "den".

Let

$$P(T_1, \dots, T_n) = \sum \alpha_{i_1 \cdots i_n} T_1^{i_1} \cdots T_n^{i_n}$$

be a polynomial with complex coefficients, and let

$$Q(T_1, \dots, T_n) = \sum \beta_{i_1 \cdots i_n} T_1^{i_1} \cdots T_n^{i_n}$$

be a polynomial with real coefficients ≥ 0. We say that Q **dominates** P and write $P \prec Q$, if $|\alpha_{(i)}| \leq \beta_{(i)}$ for all $(i) = (i_1, \dots, i_n)$. It is then immediately verified that the relation of domination is preserved under addition, multiplication, and taking partial derivatives with respect to the variables $T_1, \dots, T_n$. Thus if $P \prec Q$, then $D_i P \prec D_i Q$, where $D_i = \partial/\partial T_i$.

Lemma 2. *Let $f_1, \dots, f_n$ be functions such that the derivative $D = d/dz$ maps the ring $\mathbf{Q}[f_1, \dots, f_n]$ into itself. There exists a number C_1 having the following property. If $Q(T_1, \dots, T_n)$ is a polynomial with rational coefficients, of total degree $\leq r$, then*

$$D^m(Q(f_1, \dots, f_n)) = Q_m(f_1, \dots, f_n)$$

where $Q_m \in \mathbf{Q}[T_1, \ldots, T_n]$ is a polynomial satisfying:

(i) $\deg Q_m \ll m + r$.

(ii) size $Q_m \leq$ (size P)$m! \, C_1^{m+r}$.

(iii) There exists a denominator for the coefficients of Q_m bounded by $\text{den}(Q) C_1^{m+r}$.

Proof. Let $P_j(T_1, \ldots, T_n)$ be a polynomial such that

$$Df_j = P_j(f_1, \ldots, f_n).$$

Let d be the maximum of the degrees of $P_1, \ldots, P_n$. There exists a "differentiation" $\bar{D}$ on the polynomial ring $\mathbf{Q}[T_1, \ldots, T_n]$ such that

$$\bar{D}T_j = P_j(T_1, \ldots, T_n),$$

and for any polynomial P we have

$$\bar{D}(P(T_1, \ldots, T_n)) = \sum_{j=1}^{n} (D_j P)(T_1, \ldots, T_n) P_j(T_1, \ldots, T_n).$$

This is just obtained by the usual chain rule for differentiation, and

$$D_j = \partial/\partial T_j$$

is the usual partial derivative. But the polynomial Q is dominated by

$$Q \prec \text{size}(Q)(1 + T_1 + \cdots + T_n)^r$$

and each polynomial P_i is dominated by $\text{size}(P_i)(1 + T_1 + \cdots + T_n)^d$. Thus for some constant C_2 we have

$$\bar{D}Q \prec \text{size}(Q) C_2 r (1 + T_1 + \cdots + T_n)^{r+d}.$$

Proceeding inductively, we see that $\bar{D}^k Q$ is dominated by

$$\bar{D}^k Q \prec \text{size}(Q) C_3^k r(r + d) \cdots (r + kd)(1 + T_1 + \cdots + T_n)^{r+kd}.$$

Since

$$r(r + d) \cdots (r + kd) \leq dr(dr + d) \cdots (dr + kd)$$
$$\leq d^{k+1} r(r + 1) \cdots (r + k),$$

this product is estimated by

$$d^{k+1} \frac{(r + k)!}{r! \, k!} rk! \leq C_4^{r+k} k!.$$

This proves the lemma.

We apply the lemma when we want to evaluate a derivative

$$D^k f(w)$$

at some point w, where $f = Q(f_1, \ldots, f_n)$ is a polynomial in the functions $f_1, \ldots, f_n$. Then all we have to do is plug in $f_1(w), \ldots, f_n(w)$ in $Q_k(T_1, \ldots, T_n)$ to obtain

$$D^k f(w) = Q_k(f_1(w), \ldots, f_n(w)).$$

If w is regarded as fixed, this gives us an estimate for $D^k f(w)$ as in (ii) and (iii) of the theorem, whenever $f_1(w), \ldots, f_n(w)$ are rational numbers. Thus the previous discussion tells us how fast the denominators and absolute values of a derivative

$$D^k f(w)$$

grow when w is a point such that $f_1(w), \ldots, f_n(w)$ are rational numbers.

We now come to the main part of the proof of the theorem. Let f, g be two functions among $f_1, \ldots, f_n$ which are algebraically independent. Let L be a positive integer divisible by $2N$. We shall let L tend to infinity at the end of the proof.

Let

$$F = \sum_{i,j=1}^{L} b_{ij} f^i g^j$$

have integer coefficients, and let $L^2 = 2MN$. We wish to select the coefficients b_{ij} not all 0 such that

$$D^m F(w_\nu) = 0$$

for $m = 0, \ldots, M - 1$ and $\nu = 1, \ldots, N$. This amounts to solving a system of linear equations

$$\sum_{i,j=1}^{L} b_{ij} D^m(f^i g^j)(w_\nu) = 0,$$

and by hypothesis $D^m(f^i g^j)(w_\nu)$ is a rational number for each ν. We treat the b_{ij} as unknowns, and wish to apply Siegel's lemma. We have:

Number of unknowns $= L^2$,

Number of equations $= MN$.

Then our choice of L related to M is such that

$$\frac{\text{\# equations}}{\text{\# unknowns} - \text{\# equations}} = 1.$$

We multiply the equations by a common denominator for the coefficients. Using the estimate of Lemma 2, and Siegel's lemma, we can take the b_{ij} to be integers, whose size is bounded by

$$\text{size } b_{ij} \ll M! \, C_2^{M+L} \ll M^M C_2^{M+L}$$

for $M \to \infty$.

Since f, g are algebraically independent, the function F is not identically zero. Let s be the smallest integer such that all derivatives of F up to order $s - 1$ vanish at all points $w_1, \ldots, w_N$, but such that $D^s F$ does not vanish at one of the w_ν, say w_1. Then $s \geq M$. We let

$$\alpha = D^s F(w_1).$$

Then $\alpha \neq 0$ is a rational number, and by Lemma 2 it has a denominator which is $\ll C_1^s$ for $s \to \infty$. Let c be this denominator. Then $c\alpha$ is an integer, and its absolute value is therefore ≥ 1.

We shall obtain an upper bound for $|D^s F(w_1)|$ by the technique of Theorem 1.1. We have

$$D^s F(w_1) = s! \left. \frac{F(z)}{(z - w_1)^s} \right|_{z = w_1}.$$

We estimate the function

$$H(z) = s! \frac{F(z)}{[(z - w_1) \cdots (z - w_N)]^s} \prod_{\nu \neq 1} (w_1 - w_\nu)^s$$

on the circle of radius $R = s^{1/2\rho}$. By the maximum modulus principle, we have

$$|D^s F(w_1)| = |H(w_1)| \leq \|H\|_R \leq s^s C^{Ns} \|F\|_R / R^{Ns}$$

for a suitable constant C. Using the estimate for the coefficients of F, and the order of growth of the functions $f_1, \ldots, f_n$, together with the fact that $L \leq M \leq s$, we obtain

$$\|F\|_R \leq \frac{C_3^s s^s C_4^{R^\rho L}}{R^{Ns}} \leq \frac{C_3^s s^s C_4^{sN}}{e^{Ns(\log s)/2\rho}}.$$

Hence

$$1 \leq |c\alpha| = |cD^s F(w_1)| \leq \frac{s^{2s} C_5^{Ns}}{e^{Ns(\log s)/2\rho}}.$$

Taking logs yields

$$\frac{Ns \log s}{2\rho} \leq 2s \log s + C_6 Ns.$$

We divide by $s \log s$, and let $L \to \infty$ at the beginning of the proof, so $s \to \infty$. The inequality

$$N \leq 4\rho$$

drops out, thereby proving the theorem.

IX §4. The Phragmen–Lindelöf and Hadamard Theorems

We write a complex number in the form

$$s = \sigma + it$$

with real σ, t.

We shall use the O notation as follows. Let f, g be functions defined on a set S, and g real positive. We write

$$f(z) = O(g(z)) \qquad \text{or} \qquad |f(z)| \ll g(z) \qquad \text{for} \quad |z| \to \infty$$

if there is a constant C such that $|f(z)| \ll Cg(z)$ for $|z|$ sufficiently large. When the context makes it clear, we omit the reference that $|z| \to \infty$.

Phragmen–Lindelöf Theorem. *Let f be holomorphic in a strip $\sigma_1 \leq \sigma \leq \sigma_2$, and bounded by 1 in absolute value on the sides of the strip. Assume that there is a number $\alpha \geq 1$ such that $f(s) = O(e^{|s|^\alpha})$ in the strip. Then f is bounded by 1 in the whole strip.*

Proof. For all sufficiently large $|t|$ we have

$$|f(\sigma + it)| \leq e^{|t|^\lambda}$$

if we take $\lambda > \alpha$. Select an integer $m \equiv 2 \pmod 4$ such that $m > \lambda$. If $s = re^{i\theta}$, then

$$s^m = r^m(\cos m\theta + i \cdot \sin m\theta),$$

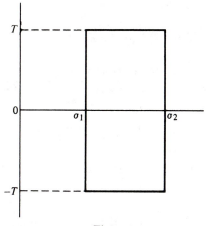

Figure 3

and $m\theta$ is close to π.   Consider the function

$$g_\epsilon(s) = g(s) = f(s)e^{\epsilon s^m},$$

with $\epsilon > 0$.   Then for s in the strip we get

$$|g(s)| \leqq e^{|t|^\lambda}e^{\epsilon r^m \cos m\theta}.$$

Consequently for large T the function $g(s)$ is bounded by 1 on the horizontal segment $t = T$, $\sigma_1 \leqq \sigma \leqq \sigma_2$.   It is also clear that $|g(s)|$ is bounded by 1 on the sides of the rectangle, as shown in Fig. 3.   Hence

$$|f(s)| \leqq e^{-\epsilon r^m \cos m\theta}$$

inside the rectangle.   This is true for every $\epsilon > 0$, and hence

$$|f(s)| \leqq 1$$

inside the rectangle, thus proving the theorem.

In the Phragmen–Lindelöf theorem we were interested in the crude asymptotic behavior for large t.   In the next theorem, we want a more refined behavior, and so we must assume that the function is holomorphic and bounded in a whole strip.

First Convexity Theorem.   *Let* $s = \sigma + it$.   *Let* f *be holomorphic and bounded on the strip* $a \leqq \sigma \leqq b$.   *For each* σ *let*

$$M_f(\sigma) = M(\sigma) = \sup_t |f(\sigma + it)|.$$

Then $\log M(\sigma)$ *is a convex function of* σ.

Proof. We must show that

$$M(\sigma)^{b-a} \leq M(a)^{b-\sigma} M(b)^{\sigma-a}.$$

We first consider the case when $M(a) = M(b) = 1$. We must show that $M(\sigma) \leq 1$. Suppose that $|f(s)| \leq B$ in the strip. For $\epsilon > 0$, let

$$g_\epsilon(s) = \frac{1}{1 + \epsilon(s - a)}.$$

Then the real part of $1 + \epsilon(s - a)$ is ≥ 1, so that $|g_\epsilon(s)| \leq 1$. Also, for $t \neq 0$,

$$|g_\epsilon(s)| \leq \frac{1}{\epsilon|t|},$$

and therefore

$$|f(s)g_\epsilon(s)| \leq \frac{B}{\epsilon|t|}.$$

Let ϵ be small, and select $t = \pm B/\epsilon$. On the boundary of the rectangle with sides at $\sigma = a$, $\sigma = b$, with top and bottom $\pm B/\epsilon$, we find that $|fg_\epsilon|$ is bounded by 1. Hence $|fg_\epsilon| \leq 1$ on the whole rectangle. Letting ϵ tend to 0 we get what we want, namely, $|f| \leq 1$ on the strip.

In general, let

$$h(s) = M(a)^{(b-s)/(b-a)} M(b)^{(s-a)/(b-a)}.$$

Then h is entire, has no zeros, and $1/h$ is bounded on the strip. We have

$$|h(a + it)| = M(a) \qquad \text{and} \qquad |h(b + it)| = M(b)$$

for all t. Consequently,

$$M_{f/h}(a) = M_{f/h}(b) = 1.$$

The first part of the proof implies that $|f/h| \leq 1$, whence $|f| \leq |h|$, thus proving our theorem.

Corollary (Hadamard Three Circle Theorem). *Let $f(z)$ be holomorphic on the annulus $\alpha \leq |z| \leq \beta$, centered at the origin. Let*

$$M(r) = \sup_{|z|=r} |f(z)|.$$

Then $\log M(r)$ *is a convex function of* $\log r$. *In other words,*

$$\log(\beta/\alpha)\log M(r) \leq \log(\beta/r)\log M(\alpha) + \log(r/\alpha)\log M(\beta).$$

Proof. Let $f^*(s) = f(e^s)$. Then f^* is holomorphic and bounded on the strip $a \leq \sigma \leq b$, where $e^a = \alpha$ and $e^b = \beta$. We simply apply the theorem, to get the corollary.

In the next corollary, we analyze a growth exponent. Let f be holomorphic in the neighborhood of a vertical line $\sigma + it$, with fixed σ, and suppose that

$$f(\sigma + it) \ll |t|^\gamma$$

for some positive number γ. The inf of all such γ can be called the **growth exponent** of f, and will be denoted by $\psi(\sigma)$. Thus

$$f(\sigma + it) \ll |t|^{\psi(\sigma) + \epsilon}$$

for every $\epsilon > 0$, and $\psi(\sigma)$ is the least exponent which makes this inequality true.

Second Convexity Theorem.   *Let f be holomorphic in the strip $a \leq \sigma \leq b$. For each σ assume that $f(\sigma + it)$ grows at most like a power of $|t|$, and let $\psi(\sigma)$ be the least number ≥ 0 for which*

$$f(\sigma + it) \ll |t|^{\psi(\sigma) + \epsilon}$$

for every $\epsilon > 0$. Assume for simplicity also that $f(\sigma + it) \ll e^{|t|^\alpha}$ in the strip, with some α, $1 \leq \alpha$. Then $\psi(\sigma)$ is convex as a function of σ, and in particular is continuous on $[a, b]$.

Proof. The Phragmen–Lindelöf theorem shows that there is a uniform M such that $f(\sigma + it) \ll |t|^M$ in the strip. Let $L_\epsilon(s)$ be the formula for the straight line segment between $\psi(a) + \epsilon$ and $\psi(b) + \epsilon$; in other words, let

$$L_\epsilon(s) = \frac{b - s}{b - a}[\psi(a) + \epsilon] + \frac{s - a}{b - a}[\psi(b) + \epsilon].$$

The function

$$f(s)(-is)^{-L_\epsilon(s)}$$

is then immediately seen to be bounded in the strip, and our theorem follows, since we get $\psi(\sigma) \leq L_\epsilon(\sigma)$ for each σ in the strip, and every $\epsilon > 0$.

Example. Define the **zeta function**

$$\zeta(s) = \sum_{n=1}^{\infty} n^{-s}.$$

It is easily shown that the series converges absolutely for $\operatorname{Re}(s) > 1$, and uniformly on every compact subset of this right half-plane, thus defining a holomorphic function. It is also easily shown that

$$\zeta(s) = \prod_{p} (1 - p^{-s})^{-1} \qquad \text{(called the \textbf{Euler product}),}$$

where the product is taken over all the prime numbers. In the next chapter you will learn about the gamma function $\Gamma(s)$. Define

$$f(s) = \pi^{-s/2} \Gamma\left(\frac{s}{2}\right) \zeta(s).$$

It can be shown that there is an extension of f to an entire function, satisfying the functional equation

$$f(s) = f(1 - s).$$

For $\operatorname{Re}(s) \geq 2$, say, the Euler product is uniformly bounded. The functional equation gives a bound for $\operatorname{Re}(s) \leq -2$. The Phragmen–Lindelöf theorem then gives a bound in the middle.

EXERCISES IX §4

Phragmen–Lindelöf for Sectors

1. Let U be the right half plane ($\operatorname{Re} z > 0$). Let f be continuous on the closure of U and analytic on U. Assume that there are constants $C > 0$ and $\alpha < 1$ such that

$$|f(z)| \leq C e^{|z|^{\alpha}}$$

for all z in U. Assume that f is bounded by 1 on the imaginary axis. Prove that f is bounded by 1 on U. Show that the assertion is not true if $\alpha = 1$.

2. More generally, let U be the open sector between two rays from the origin. Let f be continuous on the closure of U (i.e. the sector and the rays), and analytic on U. Assume that there are constants $C > 0$ and α such that

$$|f(z)| \leq C e^{|z|^{\alpha}}$$

for all $z \in U$. If π/β is the angle of the sector, assume that $0 < \alpha < \beta$. If f is bounded by 1 on the rays, prove that f is bounded by 1 on U.

(**Remark.**   The bound of 1 is used only for normalization purposes.   If f is bounded by some constant B, then dividing f by B reduces the problem to the case when the bound is 1.)

3. Let f be holomorphic on the disc D_R of radius R.   For $0 \leqq r < R$ let

$$I(r) = \frac{1}{2\pi} \int_0^{2\pi} |f(re^{i\theta})|^2 \, d\theta.$$

Let $f = \sum a_n z^n$ be the power series for f.
(a)   Show that

$$I(r) = \sum |a_n|^2 r^{2n}.$$

(b) $I(r)$ is an increasing function of r.
(c) $|f(0)|^2 \leqq I(r) \leqq \|f\|_r^2$.
(d) $\log I(r)$ is a convex function of $\log r$, assuming that f is not the zero function. [*Hint*: Put $s = \log r$,

$$J(s) = I(e^s).$$

Show that $(\log J)'' = \dfrac{J''J - (J')^2}{J^2}$.   Use the Schwarz inequality to show that

$$J''J - (J')^2 \geqq 0.]$$

IX §5. Bounds by the Real Part, Borel–Carathéodory Theorem

We shall now give a simple proof that an analytic function is essentially bounded by its real part.   As usual, we use the technique of two circles. If u is a real function, we let $\sup_R u = \sup u(z)$ for $|z| = R$.

Theorem 5.1 (Borel–Carathéodory).   *Let f be holomorphic on a closed disc of radius R, centered at the origin.   Let $\|f\|_r = \max|f(z)|$ for $|z| = r < R$.   Then*

$$\|f\|_r \leqq \frac{2r}{R - r} \sup_R \operatorname{Re} f + \frac{R + r}{R - r} |f(0)|.$$

Proof.   Let $A = \sup_R \operatorname{Re} f$.   Assume first that $f(0) = 0$.   Let

$$g(z) = \frac{f(z)}{z(2A - f(z))}.$$

Then g is holomorphic for $|z| \leqq R$.   Furthermore, if $|z| = R$, then

$$|2A - f(z)| \geqq |f(z)|.$$

Hence $\|g\|_R \leq 1/R$.   By the maximum modulus principle, we have $\|g\|_r \leq \|g\|_R$, and hence, if $|w| = r$, we get

$$\frac{|f(w)|}{r|2A - f(w)|} \leq \frac{1}{R},$$

whence

$$|f(w)| \leq \frac{r}{R}(2A + |f(w)|),$$

and therefore

$$\|f\|_r \leq \frac{2r}{R - r}A,$$

which proves the lemma in this case.

In general, we apply the preceding estimate to the function

$$h(z) = f(z) - f(0).$$

Then

$$\sup_R \operatorname{Re} h \leq \sup_R \operatorname{Re} f + |f(0)|,$$

and if $|w| = r$, we get

$$|f(w) - f(0)| \leq \frac{2r}{R - r}[A + |f(0)|],$$

whence

$$|f(w)| \leq \frac{2r}{R - r}[A + |f(0)|] + |f(0)|$$

thereby proving the theorem.

In Chapter VIII we gave a more precise description of the relationship which exists between an analytic function and its real part, by expressing the function in terms of an appropriate integral involving only the real part and the Poisson kernel function.   The analytic function is essentially uniquely determined by its real part, except for adding a constant, which explains the occurrence of the term $f(0)$ in the above estimate.   However, the proof here with the maximum modulus principle is so simple that we found it worthwhile including it any way.   Besides, we give an estimate in terms of $\sup_R \operatorname{Re} f$, not just $\|\operatorname{Re} f\|_R$.   This is significant in the next corollaries.

Corollary 5.2. *Let h be an entire function. Let $\rho > 0$. Assume that there exists a number $C > 0$ such that for all sufficiently large R we have*

$$\sup_R \operatorname{Re} h \leqq CR^\rho.$$

Then h is a polynomial of degree $\leqq \rho$.

Proof. In the Borel–Carathéodory theorem, use $R = 2r$. Then $\|h\|_r \ll r^\rho$ for $r \to \infty$. Let $h(z) = \sum a_n z^n$. By Cauchy's formula we have $|a_n| \leqq \|h\|_R/R^n$ for all R, so $a_n = 0$ if $n > \rho$, as desired.

Corollary 5.3 (Hadamard). *Let f be an entire function with no zeros. Assume that there is a constant $C \geqq 1$ such that $\|f\|_R \leqq C^{R^\rho}$ for all R sufficiently large. Then $f(z) = e^{h(z)}$ where h is a polynomial of degree $\leqq \rho$.*

Proof. Write $f = e^h$ where h is entire. The assumption implies that $\operatorname{Re} h$ satisfies the hypotheses of Corollary 5.2, whence h is a polynomial, as desired.

Entire and Meromorphic Functions

A function is said to be **entire** if it is analytic on all of C. It is said to be **meromorphic** if it is analytic except for isolated singularities which are poles. In this chapter we describe such functions more closely. We develop a multiplicative theory for entire functions, giving factorizations for them in terms of their zeros, just as a polynomial factors into linear factors determined by its zeros. We develop an additive theory for meromorphic functions, in terms of their principal part (polar part) at the poles.

X §1. Infinite Products

Lemma 1.1. *Let* $\{\alpha_n\}$ *be a sequence of complex numbers* $\alpha_n \neq 1$ *for all n. Suppose that*

$$\sum |\alpha_n|$$

converges. Then the sequence of partial products

$$P_N = \prod_{n=1}^{N} (1 - \alpha_n)$$

converges absolutely.

Proof. For all but a finite number of n, we have $|\alpha_n| < \frac{1}{2}$, so

$$\log(1 - \alpha_n)$$

is defined by the usual series, and for some constant C,

$$|\log(1 - \alpha_n)| \leq C|\alpha_n|.$$

Given ϵ, there exists N_0 such that for $N \geq N_0$ we have

$$\left| \log \prod_{n=N_0}^{N} (1 - \alpha_n) \right| \leq \prod_{n=N_0}^{N} |\alpha_n| < \epsilon.$$

Hence if we take any determination of the logarithm for a finite number of factors, and the determination given by the power series near 1 for the remaining factors, we see that the series

$$\sum_{n=1}^{N} \log(1 - \alpha_n)$$

for $N \to \infty$ converges absolutely. Since the exponential function is continuous, we can exponentiate the log of the partial products to see that

$$\prod_{n=1}^{\infty} (1 - \alpha_n) = \lim_{N \to \infty} \prod_{n=1}^{N} (1 - \alpha_n)$$

converges absolutely, as was to be shown.

The lemma reduces the study of convergence of an infinite product to the study of convergence of a series, which is more easily manageable. In the applications, we shall consider a product

$$\prod_{n=1}^{\infty} (1 - h_n(z)),$$

where $h_n(z)$ is a function $\neq 1$ for all z in a certain set K, and such that we have a bound

$$|h_n(z)| \leq |\alpha_n|$$

for all but a finite number of n, and all z in K. Taking again the partial sums of the logarithms,

$$\sum_{n=1}^{N} \log(1 - h_n(z)),$$

we can compare this with the sums of the lemma to see that these sums converge absolutely and uniformly on K. Hence the product taken with the functions $h_n(z)$ converges absolutely and uniformly for z in K.

It is useful to formulate a lemma on the logarithmic derivative of an infinite product, which will apply to those products considered above.

Lemma 1.2. *Let $\{f_n\}$ be a sequence of analytic functions on an open set U. Assume that f_n converges uniformly to 1 on U. Let $f_n(z) = 1 + h_n(z)$, and assume that the series*

$$\sum h_n(z)$$

converges uniformly and absolutely on U. Let K be a compact subset of U not containing any of the zeros of the functions f_n for all n. Then for $z \in K$ we have

$$f'/f(z) = \sum_{n=1}^{\infty} f'_n/f_n(z),$$

and the convergence is absolute and uniform on K.

Proof. By covering K with a finite number of discs of sufficiently small radius, using the compactness, we may assume that K is a closed disc. Write

$$f(z) = \prod_{n=1}^{N-1} f_n(z) \prod_{n=N}^{\infty} f_n(z),$$

where N is picked so large that $\sum |h_n(z)| < 1$. Then the series

$$\sum_{n=N}^{\infty} \log f_n(z) = \sum_{n=N}^{\infty} \log(1 + h_n(z))$$

converges uniformly and absolutely, to define a determination of

$$\log G(z), \quad \text{where} \quad G(z) = \prod_{n=N}^{\infty} f_n(z).$$

Then

$$f'/f(z) = \sum_{n=1}^{N} f'_n/f_n(z) + G'/G(z).$$

But we can differentiate the series for $\log G(z)$ term by term, whence the expression for $f'/f(z)$ follows. If the compact set is away from the zeros of f_n for all n, then the convergence is clearly uniform, as desired.

EXERCISES X §1

1. Let $0 < |\alpha| < 1$ and let $|z| \leqq r < 1$. Prove the inequality

$$\frac{\alpha + |\alpha|z}{(1 - \bar{\alpha}z)\alpha} \leqq \frac{1 + r}{1 - r}.$$

2. **(Blaschke Products).** Let $\{\alpha_n\}$ be a sequence in the unit disc D such that $\alpha_n \neq 0$ for all n, and

$$\sum_{n=1}^{\infty} (1 - |\alpha_n|)$$

converges. Show that the product

$$f(z) = \prod_{n=1}^{\infty} \frac{\alpha_n - z}{1 - \bar{\alpha}_n z} \frac{|\alpha_n|}{\alpha_n}$$

converges uniformly for $|z| \leqq r < 1$, and defines a holomorphic function on the unit disc having precisely the zeros α_n and no other zeros. Show that $|f(z)| \leqq 1$,

3. Let $\alpha_n = 1 - 1/n^2$ in the preceding exercise. Prove that

$$\lim_{x \to 1} f(x) = 0 \qquad \text{if} \quad 0 < x < 1.$$

In fact, prove the estimate for $\alpha_{n-1} < x < \alpha_n$:

$$|f(x)| < \prod_{k=1}^{n-1} \frac{x - \alpha_k}{1 - \alpha_k x} < \prod_{k=1}^{n-1} \frac{\alpha_n - \alpha_k}{1 - \alpha_k} < 2e^{-n/3}.$$

4. Prove that there exists a bounded analytic function f on the unit disc for which each point of the unit circle is a singularity.

5. **(q-Products).** Let $z = x + iy$ be a complex variable, and let $\tau = u + iv$ with u, v real, $v > 0$ be a variable in the upper half-plane H. We define

$$q_\tau = e^{2\pi i \tau} \qquad \text{and} \qquad q_z = e^{2\pi i z}.$$

Consider the infinite product

$$(1 - q_z) \prod_{n=1}^{\infty} (1 - q_\tau^n q_z)(1 - q_\tau^n / q_z).$$

(a) Prove that the infinite product is absolutely convergent.
(b) Prove that for fixed τ, the infinite product defines a holomorphic function of z, with zeros at the points

$$m + n\tau, \quad \text{with } m, n \text{ integers.}$$

We define the **second Bernoulli polynomial**

$$\mathbf{B}_2(y) = y^2 - y + \tfrac{1}{6}.$$

Define the **Néron–Green function**

$$\lambda(z, \tau) = \lambda(x, y, \tau) = -\log \left| q_\tau^{\mathbf{B}_2(y/v)/2}(1 - q_z) \prod_{n=1}^{\infty} (1 - q_\tau^n q_z)(1 - q_\tau^n/q_z) \right|.$$

(c)   Prove that for fixed τ, the function $z \mapsto \lambda(z, \tau)$ is periodic with periods $1, \tau$.

X §2. Weierstrass Products

Let f, g be entire functions with the same zeros, at which they have the same multiplicities. Then f/g is an entire function without zeros. We first analyze this case.

Theorem 2.1. *Let f be an entire function without zeros. Then there exists an entire function h such that*

$$f(z) = e^{h(z)}.$$

Proof. Let C be a closed path. Then

$$\int_C \frac{f'(z)}{f(z)} \, dz = 0$$

because f has no zeros (cf. Theorem 1.5 of Chapter VI, the residue formula). Hence the function defined by

$$h(z) = \log f(z_0) + \int_{z_0}^z \frac{f'(\zeta)}{f(\zeta)} \, d\zeta,$$

where z_0 is some fixed point in the plane, is well defined because the integral is independent of the path, and h is also analytic. Locally in the neighborhood of any point, this function is some determination of $\log f(z)$. Hence

$$e^{h(z)} = f(z),$$

as was to be shown.

We see that if f, g are two functions with the same zeros and same multiplicities, then

$$f(z) = g(z)e^{h(z)}$$

for some entire function $h(z)$. Conversely, if $h(z)$ is entire, then $g(z)e^{h(z)}$ has the same zeros as g, counted with their multiplicities.

We next try to give a standard form for a function with prescribed zeros. Suppose we order these zeros by increasing absolute value, so let $z_1, z_2, \ldots$ be a sequence of complex numbers $\neq 0$, satisfying

$$|z_1| \leq |z_2| \leq \cdots.$$

Assume that $|z_n| \to \infty$ as $n \to \infty$. If we try to define the function by the product

$$\prod_{n=1}^{\infty} \left(1 - \frac{z}{z_n}\right),$$

then we realize immediately that this product may not converge, and so we have to insert a convergence factor. We do not want this factor to introduce new zeros, so we make it an exponential. We want it to be as simple as possible, so we make it the exponential of a polynomial, whose degree will depend on the sequence of z_n. Thus we are led to consider factors of the form

$$E(z, n) = (1 - z)e^{z + z^2/2 + \cdots + z^n/n}.$$

The polynomial in the exponent is exactly what is needed to cancel the first n terms in the series for the log, so that

$$\log E(z, n) = \log(1 - z) + z + \frac{z^2}{2} + \cdots + \frac{z^n}{n}$$

$$= \sum_{k=n+1}^{\infty} \frac{-z^k}{k}.$$

Lemma 2.2. *If* $|z| \leq 1/2$, *then*

$$|\log E(z, n)| \leq 2|z|^{n+1}.$$

Proof.

$$|\log E(z, n)| \leq \frac{|z|^{n+1}}{n+1} \sum_{k=0}^{\infty} \frac{1}{2^k} \leq 2|z|^{n+1}.$$

Given the sequence $\{z_n\}$, we pick integers k_n such that the series

$$\sum_{n=1}^{\infty} \left(\frac{R}{|z_n|}\right)^{k_n}$$

converges for all positive real R. Since $|z_n| \to \infty$ we can find such k_n. We let

$$P_n(z) = z + \frac{z^2}{2} + \cdots + \frac{z^{k_n - 1}}{k_n - 1}$$

and

$$E_n(z) = \left(1 - \frac{z}{z_n}\right)e^{P_n(z/z_n)}.$$

Note that

$$P_n\left(\frac{z}{z_n}\right) = \frac{z}{z_n} + \frac{1}{2}\left(\frac{z}{z_n}\right)^2 + \cdots + \frac{1}{k_n - 1}\left(\frac{z}{z_n}\right)^{k_n - 1}.$$

Theorem 2.3. *The product*

$$\prod_{n=1}^{\infty} E_n(z) = \prod_{n=1}^{\infty}\left(1 - \frac{z}{z_n}\right)e^{P_n(z/z_n)}$$

converges uniformly and absolutely on every disc $|z| \leq R$, and defines an entire function with zeros at the points of the sequence $\{z_n\}$, and no other zeros.

Proof. Fix R. Let N be such that

$$|z_N| \leq 2R < |z_{N+1}|.$$

Then for $|z| \leq R$ and $n > N$ we have

$$\left|\frac{z}{z_n}\right| < \frac{1}{2},$$

and hence

$$|\log E_n(z)| \leq 2\left(\frac{R}{|z_n|}\right)^{k_n}.$$

Therefore the series

$$\sum_{n=1}^{\infty} \log E_n(z)$$

converges absolutely and uniformly when $|z| \leq R$, thereby implying the absolute and uniform convergence of the exponentiated product.

The limiting function obviously has the sequence $\{z_n\}$ as zeros, with the multiplicity equal to the number of times that z_n is repeated in the sequence. We still have to show that the limiting function has no other zeros. We fix some radius R and consider only $|z| \leq R$. Given ϵ there exists N_0 such that if $N \geq N_0$, then

$$\left| \log \prod_{n=N_0}^{N} E_n(z) \right| = \left| \sum_{n=N_0}^{N} \log E_n(z) \right| < \epsilon,$$

by the absolute uniform convergence of the log sequence proved previously. Hence the product

$$\prod_{N_0}^{N} E_n(z)$$

is close to 1. But

$$f(z) = \prod_{n=1}^{N_0-1} \lim_{N \to \infty} \prod_{N_0}^{N} E_n(z).$$

The first product on the right has the appropriate zeros in the disc $|z| \leq R$. The limit of the second product on the right is close to 1, and hence has no zero. This proves the theorem.

The sequence $\{z_n\}$ was assumed such that $z_n \neq 0$. Of course an entire function may have a zero at 0, and to take this into account, we have to form

$$\boxed{z^m \prod_{n=1}^{\infty} \left(1 - \frac{z}{z_n}\right) e^{P_n(z/z_n)}.}$$

This function has the same zeros as the product in Theorem 2.3, with a zero of order m at the origin in addition.

In most of the examples, one can pick $k_n - 1$ equal to a fixed integer, which in all the applications we shall find in this book is equal to 1 or 2. The chapter on elliptic functions gives examples of order 2. We now give an example of order 1.

Example. We claim that

$$\sin \pi z = \pi z \prod_{n \neq 0} \left(1 - \frac{z}{n}\right) e^{z/n},$$

and

$$\cot z = \frac{\cos z}{\sin z} = \frac{1}{z} + \sum_{n \neq 0}^{\infty} \left(\frac{1}{z - n} + \frac{1}{n}\right).$$

Proof. We let $z_n = n$ where n ranges over the integers $\neq 0$ (possibly negative). Let $k_n = 2$, $k_n - 1 = 1$. The series

$$\sum_{n=1}^{\infty} (R/n)^2$$

converges for all R. Hence the theorem implies that the function

$$f(z) = \pi z \prod_{n \neq 0} \left(1 - \frac{z}{n}\right) e^{z/n}$$

is an entire function with zeros of order 1 at the integers. *We assert that it is equal to* $\sin \pi z$. The Weierstrass product has the property stated in Lemma 1.2, and therefore taking the logarithmic derivative term by term yields

$$f'/f(z) = \frac{1}{z} + \sum_{n \neq 0} \left(\frac{1}{z - n} + \frac{1}{n}\right).$$

The convergence is uniform on every compact set not containing an integer. The function $g(z) = f'/f(z)$ has a pole of order 1 at each integer. We contend that g is periodic, with period 1. Consider the function

$$g(z + 1) - g(z),$$

and take its derivative. We obtain 0, because the derivative of both $g(z + 1)$ and $g(z)$ is

$$- \sum_{n=-\infty}^{\infty} \frac{1}{(z - n)^2}$$

Hence $g(z + 1) - g(z)$ is constant. But in this difference, the term $1/z$ has canceled, and $g(z + 1) - g(z)$ is defined at 0, with the value 0. Hence this constant is 0, and g is therefore periodic as asserted.

There is an entire function $\varphi(z)$ such that

$$f(z) = e^{\varphi(z)} \sin \pi z,$$

because $f(z)$ and $\sin \pi z$ have exactly the same zeros. Taking the logarithmic derivative yields

$$f'/f(z) = \varphi'(z) + \pi \cot \pi z,$$

where

$$\cot \pi z = \frac{\cos \pi z}{\sin \pi z} = i \frac{e^{i\pi z} + e^{-i\pi z}}{e^{i\pi z} - e^{-i\pi z}}.$$

In the strip of complex numbers $z = x + iy$ such that $0 \leq x \leq 1$, a direct substitution of $z = x + iy$ in cot πz shows that cot πz is bounded uniformly when $y \to \infty$. The same is true of $f'/f(x + iy)$, as one sees by splitting the sum

$$\sum_{n=1}^{N} \frac{1}{|x + iy - n|^2} + \sum_{N}^{\infty} \frac{1}{|x - iy - n|^2}.$$

We first take N large enough so that the second sum is small, and then take y large enough so that the first sum is small.

Consequently

$$f'/f(z) - \pi \cot \pi z = \varphi'(z)$$

is a meromorphic function, which is periodic. It has no poles, because it is immediately checked that the power series expansion of $\pi \cot \pi z$ at the origin has the same negative terms as $f'/f(z)$, whence the same negative terms at 1 by periodicity. Therefore $\varphi'(z)$ is an entire function. The estimates we have just given in the strip show that this function is bounded in the strip

$$0 \leq x \leq 1.$$

Therefore $\varphi'(z)$ is a bounded entire function, and is constant by Liouville's theorem. Putting $z = 0$ it is easy to see that $\varphi'(0) = 0$. (Check this as an exercise.) Hence $\varphi(z)$ is constant. We obtain

$$f(z) = C \sin \pi z,$$

for some constant C. We divide both sides by πz and let z tend to 0. We then see that $C = 1$, as desired.

EXERCISES X §2

1. Let f be an entire function, and n a positive integer. Show that there is an entire function g such that $g^n = f$ if and only if the orders of the zeros of f are divisible by n.

2. Let f, g be entire functions without common zeros. Show that the ideal generated by f, g in the ring of entire functions is the whole ring. In other words, show that there exist entire functions A, B such that $Af + Bg = 1$.
   [Hint: This is not trivial. One way of doing it is as follows. Given R large, write $f = F\varphi$ and $g = G\psi$, where F, G are polynomials having precisely the same zeros, counted with multiplicities, as f and g respectively on the closed disc of radius R. There exist polynomials A, B such that $AF + BG = 1$, whence

$$\frac{A}{\varphi} f + \frac{B}{\psi} g = 1.$$

Thus there exist meromorphic functions a, b, holomorphic on the closed disc D_R such that $af + bg = 1$. If a_1, b_1 is another such pair, then

$$(a - a_1)f + (b - b_1)g = 0 \qquad \text{on } D_R.$$

Since f, g have no zero in common, there exist meromorphic functions h_1, h_2, holomorphic on D_R, such that

$$a - a_1 = gh_1 \qquad \text{and} \qquad b - b_1 = fh_2,$$

whence it follows at once that $h_1 = -h$. Then prove the following inductive lemma.

Given ϵ, there exist meromorphic functions a_2, b_2 which are holomorphic on D_{2R}, such that $a_2 f + b_2 g = 1$,

$$|a_2 - a_1| \qquad \text{and} \qquad |b_2 - b_1| < \epsilon/2 \qquad \text{on } D_R.$$

To do this, solve first $a_2 f + b_2 g = 1$ on D_{2R}. Then write

$$a_2 = a_1 + hg, \qquad b_2 = b_1 - hf.$$

Let H be a polynomial which is very close to h on D_R. Replace h by $h - H$ to conclude the proof of the lemma.

The lemma will allow you to proceed stepwise to construct A, B. We leave the rest to you.]

3. Let $f_1, \ldots, f_m$ be a finite number of entire functions, and let J be the set of all combinations $A_1 f_1 + \cdots + A_m f_m$, where A_i are entire functions. Show that there exists a single entire function f such that J consists of all multiples of f, that is, J consists of all entire functions Af, where A is entire. In the language of rings, this means that every finitely generated ideal in the ring of entire functions is principal.

X §3. Functions of Finite Order

It is useful to have a simple criterion when the integers k_n in the Weierstrass product can all be taken equal to a fixed integer. Let ρ be a positive real number.

An entire function f is said to be of **order** $\leq \rho$ if, given $\epsilon > 0$, there exists a constant C (depending on ϵ) such that

$$\|f\|_R \leq C^{R^{\rho + \epsilon}} \qquad \text{for all } R \text{ sufficiently large.}$$

or equivalently,

$$\log \|f\|_R \ll R^{\rho + \epsilon} \qquad \text{for} \quad R \to \infty.$$

It is said to be of **strict order** $\leq \rho$ if the same estimate holds without the ϵ, namely

$$\|f\|_R \leq C^{R^\rho}.$$

The function is said to be of **order** ρ if ρ is the greatest lower bound of those positive numbers which make the above inequality valid, and similarly for the definition of the exact **strict order**.

Example. The function e^z has strict order 1 because $|e^z| = e^x \leq e^{|z|}$.

Theorem 3.1. *Let f be an entire function of strict order $\leq \rho$. Let $v_f(R)$ be the number of zeros of f in the disc of radius R. Then*

$$v_f(R) \ll R^\rho.$$

Proof. Dividing f by a power of z if necessary, we may assume without loss of generality that f does not vanish at the origin. By Jensen's inequality, we get

$$\int_R^{2R} \frac{v(x)}{x} \, dx \ll R^\rho.$$

But on the interval $[R, 2R]$ we have $v(x) \geq v(R)$. Hence

$$\text{left-hand side} \geq v(R) \int_R^{2R} \frac{1}{x} \, dx = v(R) \log 2,$$

This proves the theorem.

Theorem 3.2. *Let f have strict order $\leq \rho$, and let $\{z_n\}$ be the sequence of zeros $\neq 0$ of f, repeated with their multiplicities, and ordered by increasing absolute value. For every $\delta > 0$ the series*

$$\sum \frac{1}{|z_n|^{\rho+\delta}}$$

converges

Proof. We sum by parts:

$$\sum_{|z_n| \leq R} \frac{1}{|z_n|^{\rho+\delta}} \ll \sum_{r=1}^{R} \frac{v(r+1) - v(r)}{r^{\rho+\delta}}$$

$$\ll \frac{v(R)}{R^{\rho+\delta}} + \sum_{r=1}^{R} \frac{v(r)}{r^{\rho+\delta+1}} + \text{constant.}$$

Each quotient $v(r)/r^\rho$ is bounded, so the first term is bounded, and the sum is bounded by $\sum 1/r^{1+\delta}$ which converges. This proves the theorem.

Theorem 3.3 (Minimum Modulus Theorem). *Let f be an entire function of order $\leq \rho$. Let $z_1, z_2, \ldots$ be its sequence of zeros, repeated according to their multiplicities. Let $s > \rho$. Let U be the complement of the closed discs of radius $1/|z_n|^s$ centered at z_n, for $|z_n| > 1$. Then there exists $r_0(\epsilon, f)$ such that for $z \in U$, $|z| = r > r_0(\epsilon, f)$ we have*

$$|f(z)| > e^{-r^{\rho+\epsilon}}, \quad i.e. \quad \log|f(z)| > -r^{\rho+\epsilon}.$$

The theorem will first be proved for canonical products. We shall then characterize entire functions of order $\leq \rho$ by Hadamard's theorem. The general case of the minimum modulus theorem then follows at once. We now describe canonical products.

Let $\rho > 0$. Let k' be the smallest integer $> \rho$, and let $k = k' - 1$. Let $\{z_n\}$ be a sequence of complex numbers $\neq 0$, ordered by increasing absolute value, and such that

$$\sum \frac{1}{|z_n|^{\rho+\epsilon}}$$

converges for every ϵ. Let

$$P^{(k)}(z) = P(z) = z + \frac{z^2}{2} + \cdots + \frac{z^k}{k}.$$

We call

$$\boxed{E^{(k)}(z, \{z_n\}) = E(z) = \prod \left(1 - \frac{z}{z_n}\right) e^{P(z/z_n)} = \prod E\left(\frac{z}{z_n}, k\right)}$$

the **canonical product** determined by the sequence $\{z_n\}$ and the number ρ, or canonical product for short. Obvious estimates as in §2 show that it defines an entire function.

Theorem 3.4. *The above canonical product is an entire function of order $\leq \rho$.*

Proof. Let $\rho \leq \lambda \leq k + 1$. There exists a constant C such that

(1) $$|E(z, k)| \leq C^{|z|^\lambda}.$$

This is true for $|z| \leq \frac{1}{2}$ by the Lemma of Theorem 2.2, and it is even more obvious for $|z| \geq 1$ and $\frac{1}{2} \leq |z| \leq 1$. We pick $\lambda = \rho + \epsilon$ for small ϵ. Then

$$|E(z)| \leq \prod C^{|z/z_n|^\lambda} \leq C^{|z|^\lambda \Sigma 1/|z_n|^\lambda},$$

thus proving the theorem.

We now prove the minimum modulus theorem for the canonical product. Let $|z| = r$. We write

$$E(z) = \prod_{|z_n| < 2r} \left(1 - \frac{z}{z_n}\right) \prod_{|z_n| < 2r} e^{P(z/z_n)} \prod_{|z_n| > 2r} E\left(\frac{z}{z_n}, k\right).$$

The third product over $|z_n| > 2r$ is immediately seen to have absolute value $> C^{-r^{\rho+\epsilon}}$ by the usual arguments. It then follows that for some constant C',

$$\log |E(z)| \geq \sum_{|z_n| < 2r} \log\left|1 - \frac{z}{z_n}\right| - C' \sum_{|z_n| < 2r} \left|\frac{z}{z_n}\right|^k - O(r^{\rho+\epsilon}).$$

For the middle term we have the estimate

$$\sum_{|z_n| < 2r} \left|\frac{z}{z_n}\right|^k \leq r^k \sum_{m \leq 2r} \frac{v(m+1) - v(m)}{m^k}$$

$$\ll r^k \left[\frac{v(2r)}{r^k} + \sum_{m \leq 2r} \frac{v(m)}{m^{k+1}} + \text{constant}\right]$$

$$\ll r^{\rho+\epsilon}, \quad \text{by Theorem 3.1.}$$

In the first sum we have

$$\left|1 - \frac{z}{z_n}\right| = \frac{|z - z_n|}{|z_n|} \geq \frac{1}{|z_n|^{s+1}} \geq \frac{1}{(2r)^{s+1}}.$$

Hence the first sum satisfies the lower bound

$$\sum_{|z_n| < 2r} \log\left|1 - \frac{z}{z_n}\right| \geq v(2r)\log(2r)^{-s-1}$$

$$\geq -C'' r^{\rho+\epsilon} \log(2r) \quad \text{(by Theorem 3.1).}$$

which concludes the proof of the minimum modulus theorem for canonical products.

Theorem 3.5 (Hadamard). *Let f be an entire function of order ρ, and let $\{z_n\}$ be the sequence of its zeros $\neq 0$. Let $k + 1$ be the smallest integer $> \rho$. Let $P = P^{(k)}$. Then*

$$f(z) = e^{h(z)} z^m \prod \left(1 - \frac{z}{z_n}\right) e^{P(z/z_n)},$$

where m is the order of f at 0, and h is a polynomial of degree $\leq \rho$.

Proof. The series $\sum 1/|z_n|^s$ converges for $s > \rho$. Hence for every r sufficiently large, there exists R with $r \leq R \leq 2r$ such that for all n the circle of radius R does not intersect the disc of radius $1/|z_n|^s$ centered at z_n. By the minimum modulus theorem, we get a lower bound for the canonical product $E(z)$ on the circle of radius R, which shows that the quotient $f(z)/E(z)z^m$ is entire of order $\leq \rho$. We can apply Corollary 5.3 of the Borel–Carathéodory theorem to conclude the proof.

The full minimum modulus theorem for f is now obvious since the exponential term $e^{h(z)}$ obviously satisfies the desired lower bound.

X §4. Meromorphic Functions, Mittag–Leffler Theorem

Suppose f has a pole at z_0, with the power series expansion

$$f(z) = \frac{a_{-m}}{(z - z_0)^m} + \cdots + a_0 + a_1(z - z_0) + \cdots.$$

We call

$$\frac{a_{-m}}{(z - z_0)^m} + \cdots + \frac{a_{-1}}{(z - z_0)} = P\left(\frac{1}{z - z_0}\right)$$

the **principal part** of f at z_0.

We shall consider the additive analogue to the preceding section, which is to construct a meromorphic function having given principal parts at a sequence of points $\{z_n\}$ which is merely assumed to be discrete.

Theorem 4.1 (Mittag–Leffler). *Let $\{z_n\}$ be a sequence of distinct complex numbers such that $|z_n| \to \infty$. Let $\{P_n\}$ be polynomials without constant term. Then there exists a meromorphic function f whose only poles are at $\{z_n\}$ with principal part $P_n(1/(z - z_n))$. The most general function of this kind can be written in the form*

$$f(z) = \sum_n \left[P_n\left(\frac{1}{z - z_n}\right) - Q_n(z) \right] + \varphi(z),$$

where Q_n is some polynomial, and φ is entire.   The series converges absolutely and uniformly on any compact set not containing the poles.

Proof.   Since a principal part at 0 can always be added *a posteriori*, we assume without loss of generality that $z_n \neq 0$ for all n.   We expand

$$P_n\left(\frac{1}{z - z_n}\right)$$

in a power series at the origin.   We let $Q_n(z)$ be the sum of the terms of degree $\leq d_n - 1$ in this series, for a suitably chosen d_n.   Then Q_n is a polynomial of degree d_n.   For $|z| \leq |z_n|/2$ we have

$$\left| P_n\left(\frac{1}{z - z_n}\right) - Q_n(z) \right| \leq B_n \left| \frac{z}{z_n} \right|^{d_n},$$

where B_n is some fixed number depending on $P_n(1/(z - z_n))$.   We choose d_n so large that

$$\frac{B_n^{1/d_n}}{|z_n|} \to 0 \qquad \text{as} \quad |z_n| \to \infty,$$

and also such that the d_n form an increasing sequence, $d_1 < d_2 < \cdots$. Then the series

$$\sum \left[P_n\left(\frac{1}{z - z_n}\right) - Q_n(z) \right]$$

converges absolutely and uniformly for z in any compact set not containing the z_n.   In fact, given a radius R, let $R \leq |z_N|$, and split the series,

$$\sum_{n=1}^{N} \left[P_n\left(\frac{1}{z - z_n}\right) - Q_n(z) \right] + \sum_{N+1}^{\infty} \left[P_n\left(\frac{1}{z - z_n}\right) - Q_n(z) \right].$$

The first part is a finite sum.   If $|z| \leq R/2$, then the second sum is dominated by

$$\sum (B_n/|z_n|^{d_n}) z^{d_n},$$

having radius of convergence equal to ∞, as one sees by taking the d_n-th root of the coefficients.   The finite sum has the desired poles, and the infinite sum on the right has no poles in the disc of radius $R/2$.   This is true for every R, thereby completing the proof of the theorem.

CHAPTER XI

Elliptic Functions

In this chapter we give the classical example of entire and meromorphic functions of order 2. The theory illustrates most of the theorems proved so far in the book. A self-contained "analytic" continuation of the topics discussed in this chapter can be found in Chapters 3, 4, and 18 of my book on *Elliptic Functions*.

XI §1. The Liouville Theorems

By a **lattice** in the complex plane $\mathbf{C}$ we shall mean a subgroup which is free of dimension 2 over $\mathbf{Z}$, and which generates $\mathbf{C}$ over the reals. If ω_1, ω_2 is a basis of a lattice L over $\mathbf{Z}$, then we also write $L = [\omega_1, \omega_2]$. Such a lattice looks like Figure 1.

Unless otherwise specified, we also assume that $\operatorname{Im}(\omega_1/\omega_2) > 0$, i.e. that ω_1/ω_2 lies in the upper half plane $H = \{x + iy, y > 0\}$. An **elliptic function** f (with respect to L) is a meromorphic function on $\mathbf{C}$ which is L-periodic, that is,

$$f(z + \omega) = f(z)$$

for all $z \in \mathbf{C}$ and $\omega \in L$. Note that f is periodic if and only if

$$f(z + \omega_1) = f(z) = f(z + \omega_2).$$

We say that two complex numbers z, w are **congruent** mod L and we write $z \equiv w \pmod{L}$ if $z - w \in L$. Such a congruence is easily verified to be an equivalence relation. The set of equivalence classes is denoted by $\mathbf{C}/L$, and is called $\mathbf{C}$ mod L.

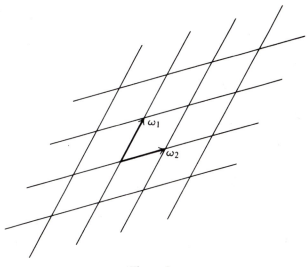

Figure 1

If $L = [\omega_1, \omega_2]$ as above, and $\alpha \in \mathbf{C}$, we call the set consisting of all points

$$\alpha + t_1\omega_1 + t_2\omega_2, \qquad 0 \leq t_i \leq 1$$

a **fundamental parallelogram** for the lattice (with respect to the given basis). We could also take the values $0 \leq t_i < 1$ to define a fundamental parallelogram, the advantage then being that in this case we get unique representatives for elements of $\mathbf{C}/L$ in $\mathbf{C}$.

An elliptic function which is entire (i.e. without poles) must be constant. Indeed, the function on a fundamental parallelogram is continuous and so bounded, and by periodicity it follows that the function is bounded on all of $\mathbf{C}$, whence constant by Liouville's theorem.

Theorem 1.1. *Let P be a fundamental parallelogram for L, and assume that the elliptic function f has no poles on its boundary ∂P. Then the sum of the residues of f in P is 0.*

Proof. We have

$$2\pi i \sum \operatorname{Res} f = \int_{\partial P} f(z)\, dz = 0,$$

this last equality being valid because of the periodicity, so the integrals on opposite sides cancel each other (Fig. 2).

Corollary. *An elliptic function has at least two poles (counting multiplicities) in $\mathbf{C}/L$.*

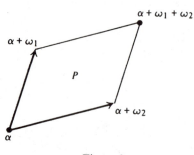

Figure 2

Theorem 1.2. *Let P be a fundamental parallelogram, and assume that the elliptic function f has no zero or pole on its boundary. Let $\{a_i\}$ be the singular points (zeros and poles) of f inside P, and let f have order m_i at a_i. Then*

$$\sum m_i = 0.$$

Proof. Observe that f elliptic implies that f' and f'/f are elliptic. We then obtain

$$0 = \int_{\partial P} f'/f(z)\, dz = 2\pi\sqrt{-1}\sum \text{Residues} = 2\pi\sqrt{-1}\sum m_i,$$

thus proving our assertion.

Theorem 1.3. *Hypotheses being as in Theorem 1.2, we have*

$$\sum m_i a_i \equiv 0 \pmod{L}.$$

Proof. This time, we take the integral

$$\int_{\partial P} z\, \frac{f'(z)}{f(z)}\, dz = 2\pi\sqrt{-1}\sum m_i a_i,$$

because

$$\text{res}_{a_i} z\, \frac{f'(z)}{f(z)} = m_i a_i.$$

On the other hand we compute the integral over the boundary of the parallelogram by taking it for two opposite sides at a time. One pair of such integrals is equal to

$$\int_{\alpha}^{\alpha+\omega_1} z\, \frac{f'(z)}{f(z)}\, dz - \int_{\alpha+\omega_2}^{\alpha+\omega_1+\omega_2} z\, \frac{f'(z)}{f(z)}\, dz.$$

We change variables in the second integral, letting $u = z - \omega_2$. Both integrals are then taken from $\alpha + \omega_1$, and after a cancellation, we get the value

$$- \omega_2 \int_\alpha^{\alpha + \omega_1} \frac{f'(u)}{f(u)} \, du = 2\pi\sqrt{-1}\, k\omega_2,$$

for some integer k. The integral over the opposite pair of sides is done in the same way, and our theorem is proved.

XI §2. The Weierstrass Function

We now prove the existence of elliptic functions by writing some analytic expression, namely the Weierstrass function

$$\wp(z) = \frac{1}{z^2} + \sum_{\omega \in L'} \left[\frac{1}{(z - \omega)^2} - \frac{1}{\omega^2} \right],$$

where the sum is taken over the set of all non-zero periods, denoted by L'. We have to show that this series converges uniformly on compact sets not including the lattice points. For bounded z, staying away from the lattice points, the expression in the brackets has the order of magnitude of $1/|\omega|^3$. Hence it suffices to prove:

Lemma 2.1. *If* $\lambda > 2$, *then* $\displaystyle\sum_{\omega \in L'} \frac{1}{|\omega|^\lambda}$ *converges.*

Proof. The partial sum for $|\omega| \leq N$ can be decomposed into a sum for ω in the annulus at n, that is, $n - 1 \leq |\omega| \leq n$, and then a sum for $1 \leq n \leq N$. In each annulus the number of lattice points has the order of magnitude n. Hence

$$\sum_{\omega \leq N} \frac{1}{|\omega|^\lambda} \ll \sum_1^\infty \frac{n}{n^\lambda} \ll \sum_1^\infty \frac{1}{n^{\lambda - 1}}$$

which converges for $\lambda > 2$.

The series expression for $\wp$ shows that it is meromorphic, with a double pole at each lattice, and no other pole. It is clear that $\wp$ is even, that is,

$$\wp(z) = \wp(-z)$$

(summing over the lattice points is the same as summing over their negatives). We get $\wp'$ by differentiating term by term,

$$\wp'(z) = -2 \sum_{\omega \in L} \frac{1}{(z - \omega)^3},$$

the sum being taken for all $\omega \in L$. Note that $\wp'$ is clearly periodic, and is odd, that is,

$$\wp'(-z) = -\wp'(z).$$

From its periodicity, we conclude that there is a constant C such that

$$\wp(z + \omega_1) = \wp(z) + C.$$

Let $z = -\omega_1/2$ (not a pole of $\wp$). We get

$$\wp\left(\frac{\omega_1}{2}\right) = \wp\left(-\frac{\omega_1}{2}\right) + C,$$

and since $\wp$ is even, it follows that $C = 0$. Hence $\wp$ is itself periodic, something which we could not see immediately from its series expansion.

Theorem 2.2. *Let f be an elliptic function periodic with respect to L. Then f can be expressed as a rational function of $\wp$ and $\wp'$.*

Proof. If f is elliptic, we can write f as a sum of an even and an odd elliptic function as usual, namely

$$f(z) = \frac{f(z) + f(-z)}{2} + \frac{f(z) - f(-z)}{2}.$$

If f is odd, then the product $f\wp'$ is even, so it will suffice to prove that if f is even, then f is a rational function of $\wp$.

Suppose that f is even and has a zero of order m at some point u. Then clearly f also has a zero of the same order at $-u$ because

$$f^{(k)}(u) = (-1)^k f^{(k)}(-u).$$

Similarly for poles.

If $u \equiv -u \,(\text{mod } L)$, then the above assertion holds in the strong sense, namely f has a zero (or pole) of even order at u.

Proof. First note that $u \equiv -u \pmod{L}$ is equivalent to

$$2u \equiv 0 \pmod{L}.$$

In $\mathbf{C}/L$ there are exactly four points with this property, represented by

$$0, \frac{\omega_1}{2}, \frac{\omega_2}{2}, \frac{\omega_1 + \omega_2}{2}$$

in a period parallelogram. If f is even, then f' is odd, that is,

$$f'(u) = -f'(-u).$$

Since $u \equiv -u \pmod{L}$ and f' is periodic, it follows that $f'(u) = 0$, so that f has a zero of order at least 2 at u. If $u \not\equiv 0 \pmod{L}$, then the above argument shows that the function

$$g(z) = \wp(z) - \wp(u)$$

has a zero of order at least 2 (hence exactly 2 by Theorem 1.2 and the fact that $\wp$ has only one pole of order 2 on the torus). Then f/g is even, elliptic, holomorphic at u. If $f(u)/g(u) \neq 0$, then $\operatorname{ord}_u f = 2$. If $f(u)/g(u) = 0$, then f/g again has a zero of order at least 2 at u and we can repeat the argument. If $u \equiv 0 \pmod{L}$ we use $g = 1/\wp$ and argue similarly, thus proving that f has a zero of even order at u.

Now let $u_i \, (i = 1, \ldots, r)$ be a family of points containing one representative from each class $(u, -u) \pmod{L}$ where f has a zero or pole, other than the class of L itself. Let

$$m_i = \operatorname{ord}_{u_i} f \qquad \text{if} \quad 2u_i \not\equiv 0 \pmod{L},$$

$$m_i = \tfrac{1}{2} \operatorname{ord}_{u_i} f \qquad \text{if} \quad 2u_i \equiv 0 \pmod{L}.$$

Our previous remarks show that for $a \in \mathbf{C}$, $a \not\equiv 0 \pmod{L}$, the function $\wp(z) - \wp(a)$ has a zero of order 2 at a if and only if $2a \equiv 0 \pmod{L}$, and has distinct zeros of order 1 at a and $-a$ otherwise. Hence for all $z \not\equiv 0 \pmod{L}$ the function

$$\prod_{i=1}^{r} [\wp(z) - \wp(u_i)]^{m_i}$$

has the same order at z as f. This is also true at the origin because of Theorem 1.2 applied to f and the above product. The quotient of the above product by f is then an elliptic function without zero or pole, hence a constant, thereby proving Theorem 2.9.

Next, we obtain the power series development of $\wp$ and $\wp'$ at the origin, from which we shall get the algebraic relation holding between these two functions. We do this by brute force.

$$\wp(z) = \frac{1}{z^2} + \sum_{\omega \in L'} \left[\frac{1}{\omega^2} \left(1 + \frac{z}{\omega} + \left(\frac{z}{\omega} \right)^2 + \cdots \right)^2 - \frac{1}{\omega^2} \right]$$

$$= \frac{1}{z^2} + \sum_{\omega \in L'} \sum_{m=1}^{\infty} (m + 1) \left(\frac{z}{\omega} \right)^m \frac{1}{\omega^2}$$

$$= \frac{1}{z^2} + \sum_{m=1}^{\infty} c_m z^m,$$

where

$$c_m = \sum_{\omega \neq 0} \frac{m + 1}{\omega^{m+2}}.$$

Note that $c_m = 0$ if m is odd.

Using the notation

$$s_m(L) = s_m = \sum_{\omega \neq 0} \frac{1}{\omega^m}$$

we get the expansion

$$\wp(z) = \frac{1}{z^2} + \sum_{n=1}^{\infty} (2n + 1)s_{2n+2}(L)z^{2n},$$

from which we write down the first few terms explicitly:

$$\wp(z) = \frac{1}{z^2} + 3s_4 z^2 + 5s_6 z^4 + \cdots$$

and differentiating term by term,

$$\wp'(z) = \frac{-2}{z^3} + 6s_4 z + 20s_6 z^3 + \cdots.$$

Theorem 2.3. *Let* $g_2 = g_2(L) = 60s_4$ *and* $g_3 = g_3(L) = 140s_6$. *Then*

$$\wp'^2 = 4\wp^3 - g_2 \wp - g_3.$$

Proof. We expand out the function

$$\varphi(z) = \wp'(z^2)^2 - 4\wp(z)^3 + g_2\wp(z) + g_3$$

at the origin, paying attention only to the polar term and the constant term. This is easily done, and one sees that there is enough cancellation so that these terms are 0, in other words, $\varphi(z)$ is an elliptic function without poles, and with a zero at the origin. Hence φ is identically zero, thereby proving our theorem.

The preceding theorem shows that the points $(\wp(z), \wp'(z))$ lie on the curve defined by the equation

$$y^2 = 4x^3 - g_2 x - g_3.$$

The cubic polynomial on the right-hand side has a discriminant given by

$$\Delta = g_2^3 - 27g_3^2.$$

We shall see in a moment that this discriminant does not vanish.
    Let

$$e_i = \wp\left(\frac{\omega_i}{2}\right), \qquad i = 1, 2, 3,$$

where $L = [\omega_1, \omega_2]$ and $\omega_3 = \omega_1 + \omega_2$. Then the function

$$h(z) = \wp(z) - e_i$$

has a zero at $\omega_i/2$, which is of even order so that $\wp'(\omega_i/2) = 0$ for $i = 1, 2, 3$, by previous remarks. Comparing zeros and poles, we conclude that

$$\wp'(z) = 4(\wp(z) - e_1)(\wp(z) - e_2)((z) - e_3).$$

Thus e_1, e_2, e_3 are the roots of $4x^3 - g_2 x - g_3$. Furthermore, $\wp$ takes on the value e_i with multiplicity 2 and has only one pole of order 2 mod L, so that $e_i \neq e_j$ for $i \neq j$. This means that the three roots of the cubic polynomial are distinct, and therefore

$$\Delta = g_2^3 - 27g_3^2 \neq 0.$$

XI §3. The Addition Theorem

Given complex numbers g_2, g_3 such that $g_2^3 - 27g_3^2 \neq 0$, one can ask whether there exists a lattice for which these are the invariants associated to the lattice as in the preceding section. The answer is yes. For the moment, we consider the case when g_2, g_3 are given as in the preceding section, that is, $g_2 = 60s_4$ and $g_3 = 140s_6$.

We have seen that the map

$$z \mapsto (\wp(z), \wp'(z))$$

parametrizes points on the cubic curve defined by the equation

$$y^2 = 4x^3 - g_2 x - g_3.$$

If $z \notin L$ then the image of z under this map is a point of the curve, and if $z \in L$, then we can define its image to be a "point at infinity". If $z_1 \equiv z_2 \pmod{L}$ then z_1 and z_2 have the same image under this map.
Let

$$P_1 = (\wp(u_1), \wp'(u_1)) \quad \text{and} \quad P_2 = (\wp(u_2), \wp'(u_2))$$

be two points on the curve. Let $u_3 = u_1 + u_2$. Let

$$P_3(\wp(u_3), \wp'(u_3)).$$

We also write

$$P_1 = (x_1, y_1), \qquad P_2 = (x_2, y_2), \qquad P_3 = (x_3, y_3).$$

Then we shall express x_3, y_3 as rational functions of (x_1, y_1) and (x_2, y_2). We shall see that P_3 is obtained by taking the line through P_1, P_2, intersecting it with the curve, and reflecting the point of intersection through the x-axis, as shown on Fig. 3.

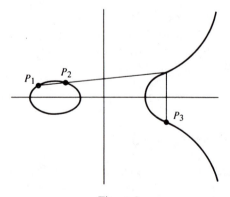

Figure 3

Select u_1, $u_2 \in \mathbf{C}$ and $\notin L$, and assume $u_1 \not\equiv u_2 \pmod{L}$. Let a, b be complex numbers such that

$$\wp'(u_1) = a\wp(u_1) + b,$$
$$\wp'(u_2) = a\wp(u_2) + b,$$

in other words $y = ax + b$ is the line through $(\wp(u_1), \wp'(u_1))$ and $(\wp(u_2), \wp'(u_2))$. Then

$$\wp'(z) - (a\wp(z) + b)$$

has a pole of order 3 at 0, whence it has three zeros, counting multiplicities, and two of these are at u_1 and u_2. If, say, u_1 had multiplicity 2, then by Theorem 1.3 we would have

$$2u_1 + u_2 \equiv 0 \pmod{L}.$$

If we fix u_1, this can hold for only one value of u_2. Let us assume that we do not deal with this value. Then both u_1, u_2 have multiplicity 1, and the third zero lies at

$$u_3 \equiv -(u_1 + u_2) \pmod{L}$$

again by Theorem 1.3. So we also get

$$\wp'(u_3) = a\wp(u_3) + b.$$

The equation

$$4x^3 - g_2 x - g_3 - (ax + b)^2 = 0$$

has three roots, counting multiplicities. They are $\wp(u_1)$, $\wp(u_2)$, $\wp(u_3)$, and the left-hand side factors as

$$4(x - \wp(u_1))(x - \wp(u_2))(x - \wp(u_3)).$$

Comparing the coefficient of x^2 yields

$$\wp(u_1) + \wp(u_2) + \wp(u_3) = \frac{a^2}{4}.$$

But from our original equations for a and b, we have

$$a(\wp(u_1) - \wp(u_2)) = \wp'(u_1) - \wp'(u_2).$$

Therefore from

$$\wp(u_3) = \wp(-(u_1 + u_2)) = \wp(u_1 + u_2)$$

we get

$$\wp(u_1 + u_2) = -\wp(u_1) - \wp(u_2) + \frac{1}{4}\left(\frac{\wp'(u_1) - \wp'(u_2)}{\wp(u_1) - \wp(u_2)}\right)^2$$

or in algebraic terms,

$$x_3 = -x_1 - x_2 + \frac{1}{4}\left(\frac{y_1 - y_2}{x_1 - x}\right)^2.$$

Fixing u_1, the above formula is true for all but a finite number of $u_2 \equiv u_1 \pmod{L}$, whence for all $u_2 \not\equiv u_1 \pmod{L}$ by analytic continuation.

For $u_1 \equiv u_2 \pmod{L}$ we take the limit as $u_1 \to u_2$ and get

$$\wp(2u) = -2\wp(u) + \frac{1}{4}\left(\frac{\wp''(u)}{\wp(u)}\right)^2.$$

These give us the desired algebraic addition formulas. Note that the formulas involve only g_2, g_3 as coefficients in the rational functions.

XI §4. The Sigma and Zeta Functions

Both in number theory and analysis one factorizes elements into prime powers. In analysis, this means that a function gets factored into an infinite product corresponding to its zeros and poles. Taking the values at special points, such an analytic expression reflects itself into special properties of the values, for which it becomes possible to determine the prime factorization in number fields.

In this chapter, we are concerned with the analytic expressions.

Our first task is to give a universal gadget allowing us to factorize an elliptic function, with a numerator and denominator which are entire functions, and are as periodic as possible.

One defines a **theta function** (on $\mathbf{C}$) with respect to a lattice L, to be an entire function θ satisfying the condition

$$\theta(z + u) = \theta(z)e^{2\pi i[l(z, u) + c(u)]}, \qquad z \in \mathbf{C}, \quad u \in L,$$

where l is **C**-linear in z, **R**-linear in u, and $c(u)$ is some function depending only on u. We shall construct a theta function.

We write down the **Weierstrass sigma function**, which has zeros of order 1 at all lattice points, by the Weierstrass product

$$\sigma(z) = z \prod_{\omega \in L'} \left(1 - \frac{z}{\omega} \right) e^{z/\omega + 1/2(z/\omega)^2}.$$

Here L' means the lattice from which 0 is deleted, i.e. we are taking the product over the non-zero periods. We note that σ also depends on L, and so we write $\sigma(z, L)$, which is homogeneous of degree 1, namely

$$\boxed{\sigma(\lambda z, \lambda L) = \lambda \sigma(z, L),} \quad \lambda \in \mathbf{C}, \quad \lambda \neq 0.$$

Taking the logarithmic derivative formally yields the **Weierstrass zeta function**

$$\zeta(z, L) = \zeta(z) = \frac{\sigma'(z)}{\sigma(z)} = \frac{1}{z} + \sum_{\omega \in L'} \left[\frac{1}{z - \omega} + \frac{1}{\omega} + \frac{z}{\omega^2} \right].$$

It is clear that the sum on the right converges absolutely and uniformly for z in a compact set not containing any lattice point, and hence integrating and exponentiating shows that the infinite product for $\sigma(z)$ also converges absolutely and uniformly in such a region. Differentiating $\zeta(z)$ term by term shows that

$$\zeta'(z) = -\wp(z) = -\frac{1}{z^2} - \sum_{\omega \in L'} \left[\frac{1}{(z - \omega)^2} - \frac{1}{\omega^2} \right].$$

Also from the product and sum expressions, we see at once that *both σ and ζ are odd functions*, that is,

$$\sigma(-z) = -\sigma(z) \quad \text{and} \quad \zeta(-z) = -\zeta(z).$$

The series defining $\zeta(z, L)$ shows that it is homogeneous of degree -1, that is,

$$\boxed{\zeta(\lambda z, \lambda L) = \frac{1}{\lambda} \zeta(z, L).}$$

Differentiating the function $\zeta(z + \omega) - \zeta(z)$ for any $\omega \in L$ yields 0 because the $\wp$-function is periodic. Hence there is a constant $\eta(\omega)$ (sometimes written η_ω) such that

$$\zeta(z + \omega) = \zeta(z) + \eta(\omega).$$

It is clear that $\eta(\omega)$ is **Z**-linear in ω. If $L = [\omega_1, \omega_2]$, then one uses the notation

$$\eta(\omega_1) = \eta_1 \quad \text{and} \quad \eta(\omega_2) = \eta_2.$$

As with ζ, the form $\eta(\omega)$ satisfies the homogeneity relation of degree -1, as one verifies directly from the similar relation for ζ. Observe that the lattice should strictly be in the notation, so that in full, the relations should read:

$$\zeta(z + \omega, L) = \zeta(z, L) + \eta(\omega, L),$$

$$\eta(\lambda\omega, \lambda L) = \frac{1}{\lambda} \eta(\omega, L).$$

Theorem 4.1. *The function σ is a theta function, and in fact*

$$\frac{\sigma(z + \omega)}{\sigma(z)} = \psi(\omega)e^{\eta(\omega)(z + \omega/2)},$$

where

$$\psi(\omega) = 1 \quad \text{if} \quad \omega/2 \in L,$$

$$\psi(\omega) = -1 \quad \text{if} \quad \omega/2 \notin L.$$

Proof. We have

$$\frac{d}{dz} \log \frac{\sigma(z + \omega)}{\sigma(z)} = \eta(\omega).$$

Hence

$$\log \frac{\sigma(z + \omega)}{\sigma(z)} = \eta(\omega)z + c(\omega),$$

whence exponentiating yields

$$\sigma(z + \omega) = \sigma(z)e^{\eta(\omega)z + c(\omega)},$$

which shows that σ is a theta function. We write the quotient as in the statement of the theorem, thereby defining $\psi(\omega)$, and it is then easy to determine $\psi(\omega)$ as follows.

Suppose that $\omega/2$ is not a period. Set $z = -\omega/2$ in the above relation. We see at once that $\psi(\omega) = -1$ because σ is odd. On the other hand, consider

$$\frac{\sigma(z + 2\omega)}{\sigma(z)} = \frac{\sigma(z + 2\omega)}{\sigma(z + \omega)} \frac{\sigma(z + \omega)}{\sigma(z)}.$$

Using the functional equation twice and comparing the two sides, we see that $\psi(2\omega) = \psi(\omega)^2$. In particular, if $\omega/2 \in L$, then

$$\psi(\omega) = \psi(\omega/2)^2.$$

Dividing by 2 until we get some element of the lattice which is not equal to twice a period, we conclude at once that $\psi(\omega) = (-1)^{2n} = 1$.

The numbers η_1 and η_2 are called **basic quasi periods of** ζ.

Legendre Relation. *We have*

$$\eta_2 \omega_1 - \eta_1 \omega_2 = 2\pi i.$$

Proof. We integrate around a fundamental parallelogram P, just as we did for the $\wp$-function:

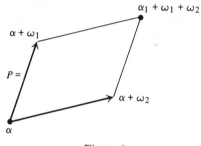

Figure 4

The integral is equal to

$$\int_{\partial P} \zeta(z) \, dz = 2\pi i \sum \text{residues of } \zeta$$

$$= 2\pi i$$

because ζ has residue 1 at 0 and no other pole in a fundamental parallelogram containing 0. On the other hand, using the quasi periodicity, the integrals over opposite sides combine to give

$$\eta_2 \omega_1 - \eta_1 \omega_2,$$

as desired.

Next, we show how the sigma function can be used to factorize elliptic functions. We know that the sum of the zeros and poles of an elliptic function must be congruent to zero modulo the lattice. Selecting suitable representatives of these zeros and poles, we can always make the sum equal to 0.

For any $a \in \mathbf{C}$ we have

$$\frac{\sigma(z + a + \omega)}{\sigma(z + a)} = \psi(\omega)e^{\eta(\omega)(z + \omega/2)}e^{\eta(\omega)a}.$$

Observe how the term $\eta(\omega)a$ occurs linearly in the exponent. It follows that if $\{a_i\}$, $\{b_i\}$ $(i = 1, \ldots, n)$ are families of complex numbers such that

$$\sum a_i = \sum b_i,$$

then the function

$$\frac{\prod \sigma(z - a_i)}{\prod \sigma(z - b_i)}$$

is periodic with respect to our lattice, and is therefore an elliptic function. Conversely, any elliptic function can be so factored into a numerator and denominator involving the sigma function. We write down explicitly the special case with the $\wp$-function.

Theorem 4.2. *For any $a \in \mathbf{C}$ not in L, we have*

$$\wp(z) - \wp(a) = -\frac{\sigma(z + a)\sigma(z - a)}{\sigma^2(z)\sigma^2(a)}.$$

Proof. The function $\wp(z) - \wp(a)$ has zeros at a and $-a$, and has a double pole at 0. Hence

$$\wp(z) - \wp(a) = C\frac{\sigma(z + a)\sigma(z - a)}{\sigma^2(z)}$$

for some constant C. Multiply by z^2 and let $z \to 0$. Then $\sigma^2(z)/z^2$ tends to 1 and $z^2\wp(z)$ tends to 1. Hence we get the value $C = -1/\sigma^2(a)$, thus proving our theorem.

CHAPTER XII

Differentiating Under an Integral

So far, we have not really used Goursat's theorem that an analytic function is holomorphic. We now come to a situation where the natural way to define a function is not through a power series but through an integral depending on a parameter.

We shall be integrating over intervals. For concreteness let us assume that we integrate on $[0, \infty[$. A function f on this interval is said to be **absolutely integrable** if

$$\int_0^\infty |f(t)| \, dt$$

exists. If the function is continuous, the integral is of course defined as the limit

$$\lim_{B \to \infty} \int_0^B |f(t)| \, dt.$$

We shall also deal with integrals depending on a parameter. This means f is a function of two variables, $f(t, z)$, where z lies in some domain U in the complex numbers. The integral

$$\int_0^\infty f(t, z) \, dt$$

is said to be **uniformly convergent** for $z \in U$ if, given ϵ, there exists B_0 such that if $B_0 < B_1 < B_2$, then

$$\left| \int_{B_1}^{B_2} f(t, z) \, dt \right| < \epsilon.$$

The integral is **absolutely and uniformly convergent** for $z \in U$ if this same condition holds with $f(t, z)$ replaced by the absolute value $|f(t, z)|$.

XII §1. The Differentiation Lemma

Lemma 1.1. *Let I be an interval of real numbers, possibly infinite. Let U be an open set of complex numbers. Let $f = f(t, s)$ be a continuous function on $I \times U$. Assume:*

(i) *For each compact subset K of U the integral*

$$\int_I f(t, s)\, dt$$

is uniformly convergent for $s \in K$.

(ii) *For each t the function $s \mapsto f(t, s)$ is analytic. Let*

$$F(s) = \int_I f(t, s)\, dt.$$

Then F is analytic on U, and

$$F'(s) = \int_I D_2 f(t, s)\, dt.$$

Furthermore $D_2 f(t, s)$ satisfies the same hypotheses as f.

Proof. Let $\{I_n\}$ be a sequence of finite closed intervals, increasing to I. Let D be a disc in the s-plane whose closure is contained in U. Let γ be the circle bounding D. Then for each s in D we have

$$f(t, s) = \frac{1}{2\pi i} \int_\gamma \frac{f(t, \zeta)}{\zeta - s}\, d\zeta,$$

so

$$F(s) = \frac{1}{2\pi i} \int_I \int_\gamma \frac{f(t, \zeta)}{\zeta - s}\, d\zeta\, dt.$$

If γ has radius R, center s_0, consider only s such that $|s - s_0| \leq R/2$. Then

$$\left| \frac{1}{\zeta - s} \right| \geq 2/R.$$

For each n we can define

$$F_n(s) = \frac{1}{2\pi i} \int_{I_n} \int_\gamma \frac{f(t, \zeta)}{\zeta - s} \, d\zeta \, dt.$$

In view of the restriction on s above, we may interchange the integrals and get

$$F_n(s) = \frac{1}{2\pi i} \int_\gamma \frac{1}{\zeta - s} \left[\int_{I_n} f(t, \zeta) \, dt \right] d\zeta.$$

Then F_n is analytic by Theorem 1.2 of Chapter V. By hypothesis, the integrals over I_n converge uniformly to the integral over I. Hence F is analytic, being the uniform limit of the functions F_n for $|s - s_0| \leqq R/2$. On the other hand, $F'_n(s)$ is obtained by differentiating under the integral sign in the usual way, and converges uniformly to $F'(s)$. However

$$F'_n(s) = \frac{1}{2\pi i} \int_{I_n} D_2 f(t, s) \, dt.$$

This proves the theorem.

Observe that the hypotheses under which the theorem is proved are slightly weaker than in the real case, because of the peculiar nature of complex differentiable functions, whose derivative can be expressed as as integral. For the differentiation lemma in the real case, cf. for instance my *Real Analysis*, Chapter XIV, §4.

Example. Let f be a continuous function with compact support on the real numbers. (**Compact support** means that the function is equal to 0 outside a compact set.) Consider the integral

$$F(z) = \int_{-\infty}^\infty f(t) e^{itz} \, dt.$$

Let $y_0 < y_1$ be real numbers. The integrand for

$$y_0 < \operatorname{Im} z < y_1$$

is of the form

$$f(t) e^{itz} = f(t) e^{itx} e^{-ty}$$

and we have $|e^{itx}| = 1$, whereas e^{-ty} lies between e^{-ty_1} and e^{-ty_0}. Since f has compact support, the values of t for which $f(t) \neq 0$ are bounded.

Hence e^{itz} is bounded uniformly for $y_0 < \text{Im } z < y_1$. Thus the integral converges absolutely and uniformly for z satisfying these inequalities.

Differentiating under the integral sign yields

$$itf(t)e^{itz},$$

and again the function $itf(t)$ has compact support. Thus the same argument can be applied. In fact we have the uniform bound

$$|itf(t)e^{itz}| \leq |t| f(t)e^{ty_1}$$

which is independent of z in the given regions. Therefore we obtain

$$F'(z) = \int_{-\infty}^{\infty} itf(t)e^{itz} \, dt \qquad \text{for} \quad y_0 < \text{Im } z < y_1.$$

As this is true for every choice of y_0, y_1 we conclude that in fact F is an entire function.

EXERCISES XII §1

1. Prove the continuity lemma:

Continuity Lemma. *Let I be an interval, U an open set in the complex numbers, and $f(t, z)$ a continuous function on $I \times U$. Assume that there exists a function φ on I which is absolutely integrable on I, and such that for all $z \in U$ we have*

$$|f(t, z)| \leq \varphi(t).$$

Then the function F defined by

$$F(z) = \int_I f(t, z) \, dt$$

is continuous.

This is easier than the differentiation lemma. Cf. my *Real Analysis*, Chapter XIV.

The Laplace Transform

2. Let f be a continuous function with compact support on the interval $[0, \infty[$. Show that the function Lf given by

$$Lf(z) = \int_0^{\infty} f(t)e^{-zt} \, dt$$

is entire.

3. Let f be a continuous function on $[0, \infty[$, and assume that there is constant $C > 1$ such that

$$|f(t)| \ll C^t \qquad \text{for} \quad t \to \infty,$$

i.e. there exist constants A, B such that $|f(t)| \leq Ae^{Bt}$ for all t sufficiently large.
(a) Prove that the function

$$Lf(z) = \int_0^\infty f(t)e^{-zt}\, dt$$

is analytic in some half plane $\operatorname{Re} z \geq \sigma$ for some real number σ. In fact, the integral converges absolutely for some σ. Either such σ have no lower bound, in which case Lf is entire, or the greatest lower bound σ_0 is called the **abscissa of convergence of the integral**, and the function Lf is analytic for $\operatorname{Re}(z) > \sigma_0$. The integral converges absolutely for

$$\operatorname{Re} z \geq \sigma_0 + \epsilon,$$

for every $\epsilon > 0$.
The function Lf is called the **Laplace transform** of f.
(b) Assuming that f is of class C^1, prove by integrating by parts that

$$(Lf)'(z) = zLf(z) - f(0).$$

Find the Laplace transform of the following functions, and the abscissa of convergence of the integral defining the transform. In each case, a is a real number $\neq 0$.

4. $f(t) = e^{-at}$ 5. $f(t) = \cos at$

6. $f(t) = \sin at$ 7. $f(t) = (e^t + e^{-t})/2$

8. Suppose that f is periodic with period $a > 0$, that is $f(t + a) = f(t)$ for all $t \geq 0$. Show that

$$Lf(z) = \frac{\int_0^a e^{-zt}f(t)\, dt}{1 - e^{-az}} \qquad \text{for} \quad \operatorname{Re} z > 0.$$

XII §2. The Gamma Function

We define

Γ 1. $$\Gamma(z) = \int_0^\infty e^{-t}t^z\, \frac{dt}{t}.$$

Then the integral fits the hypotheses of the basic lemma, provided

$$0 < a \leq \operatorname{Re} z \leq b \qquad \text{if} \quad 0 < a < b \text{ are real numbers.}$$

Therefore $\Gamma(z)$ is an analytic function in the right half plane. (Give the details of the proof that the above integral satisfies the hypotheses of the lemma, and write down explicitly what the derivative is.)

Integrals like the above are called **Mellin transforms**. We write dt/t because this expression is invariant under "multiplicative translations". This means: Let f be any function which is absolutely integrable on $0 < t < \infty$. Let a be a positive number. Then

$$\int_0^\infty f(at) \frac{dt}{t} = \int_0^\infty f(t) \frac{dt}{t}.$$

This is verified trivially by the change of variables formula of freshman calculus. Use will be made of this in the exercises. For example, replacing t by nt where n is a positive integer, we obtain

$$\frac{1}{n^s} = \int_0^\infty e^{-nt} t^s \frac{dt}{t} \qquad \text{for} \quad \text{Re}(s) > 0.$$

Summing over n yields what is called the Riemann zeta function of the complex variable s.

We now go on with the general theory of the gamma function.

Integrating by parts, the reader will prove easily that for x real > 0,

$$\int_0^n \left(1 - \frac{t}{n}\right)^n t^{x-1} \, dt = \frac{n^x n!}{x(x+1) \cdots (x+n)}$$

for every integer $n \geq 1$. To take the limit as $n \to \infty$, one then proves the inequalities

$$e^{-1}\left(1 - \frac{e}{2n} t^2\right) \leq \left(1 - \frac{t}{n}\right)^n \leq e^{-t} \qquad \text{for} \quad 0 \leq t \leq n.$$

[*Hint*: Take the log. This is freshman calculus.] It follows that we can take the limit on the left, and obtain for x real > 0:

$$\lim_{n \to \infty} \int_0^n \left(1 - \frac{t}{n}\right)^n t^{x-1} \, dt = \Gamma(x),$$

so that if we put

$$g_n(z) = \frac{z(z+1) \cdots (z+n)}{n! \, n^z}$$

then $\Gamma(z) = \lim\limits_{n \to \infty} 1/g_n(z)$, or in other words,

Γ 2.
$$\Gamma(z) = \lim_{n \to \infty} \frac{n!\, n^z}{z(z+1) \cdots (z+n)}$$

whenever z is real > 0. It will follow from the subsequent estimates that this limit is also valid for $z \neq$ negative integer.

We shall now prove that $1/\Gamma(z)$ is an entire function, and get its Weierstrass product,

Γ 3.
$$1/\Gamma(z) = ze^{\gamma z} \prod_{n=1}^{\infty} \left(1 + \frac{z}{n}\right) e^{-z/n},$$

where γ will be some constant, which will be obtained as some limit.

The functions $g_n(z)$ are entire. Furthermore, they already have a multiplicative structure, so it is reasonable that we can take their limits easily to get the Weierstrass product. We do this as follows.

For $n \geq 2$ we have

$$f_n(z) = \frac{g^n(z)}{g^{n-1}(z)} = \left(1 + \frac{z}{n}\right)\left(1 - \frac{1}{n}\right)^z.$$

Fix R. Then for $|z| \leq R$, and all n sufficiently large, one deduces at once the estimate

$$|\log f_n(z)| \leq 2R^2/n^2.$$

Therefore the general theory of products applies, and the infinite product

$$g_1 \prod_{n=2}^{\infty} g_n/g_{n-1} = g_1 \prod_{n=2}^{\infty} f_n$$

converges uniformly for $|z| \leq R$, and defines an entire function $f(z)$, whose zeros are at the negative integers,

$$n = -1, -2, \cdots.$$

From the definition of $g_n(z)$, we get directly

$$g_n(z) = z \prod_{k=1}^{n} \left(\left(1 + \frac{z}{k}\right) e^{-z/k}\right) e^{z(1 + (1/2) + \cdots + (1/n) - \log n)}.$$

From calculus, one knows that the limit

$$\lim\left(1 + \frac{1}{2} + \cdots + \frac{1}{n} - \log n\right)$$

exists, and is equal to what is known as the **Euler constant** γ. Hence we find

$$g(z) = ze^{\gamma z} \prod_{k=1}^{\infty}\left(\left(1 + \frac{z}{k}\right)e^{-z/k}\right),$$

which is the desired Weierstrass product for the limit function g.

Finally, $1/\Gamma(z)$ and $g(z)$ are analytic on the right half plane, and coincide on the real right axis. Hence they coincide on the right half plane. Therefore g is the unique entire function which coincides with $1/\Gamma(z)$ for $\mathrm{Re}(z) > 0$. This proves what we wanted.

Given the above, it is an easy matter to get other classical identities for the gamma function, for example,

Γ 4. $\Gamma(z + 1) = z\Gamma(z)$ *and* $\Gamma(1) = 1.$

We leave this as an exercise to the reader. In particular

$$\Gamma(n + 1) = n!.$$

Also prove that

Γ 5. $\Gamma(z)\Gamma(1 - z) = \dfrac{\pi}{\sin \pi z}.$

Note that the Weierstrass product for $1/\Gamma(z)$ is made up of half the zeros of $\sin \pi z$.

We also have the following relation for every positive integer N, which is sometimes called the **multiplication formula**, or **distribution relation**. Define

$$G(z) = \frac{1}{\sqrt{2\pi}} \Gamma(z).$$

Then

Γ 6. $\displaystyle\prod_{j=0}^{N-1} G\left(z + \frac{j}{N}\right) = G(Nz)N^{1/2 - Nz}.$

We leave the proof as an exercise. (Compare the zeros and poles of each side, and determine the quotient by taking a suitable limit.)

These are the principal theorems about the gamma function, except for one important but more technical result, the **Stirling formula**, which describes the asymptotic development.   The best statement is the one giving an exact error term:

Γ 7.
$$\log \Gamma(z) = \left(z - \frac{1}{2}\right)\log z - z + \tfrac{1}{2}\log(2\pi) - \int_0^\infty \frac{P_1(t)}{z + t}\, dt.$$

where $P_1(t) = t - [t] - \frac{1}{2}$ is the sawtooth function, $[t]$ denoting the largest integer $\leq t$. One takes the principal value for the log, deleting the negative real axis where the gamma function has its poles.   The usefulness of the error term involving the integral of $P_1(t)$ is that it tends to 0 uniformly in every sector of complex numbers $z = re^{i\theta}$ such that

$$-\pi + \delta \leq \theta \leq \pi - \delta, \qquad 0 < \delta < \pi.$$

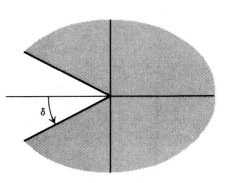

Figure 1

When $z = n$ is a positive integer, it is at the level of calculus to prove that

$$n! = n^n e^{-n}\sqrt{2\pi n}\; e^{\lambda/12n},$$

where $|\lambda| \leq 1$. Since $n! = \Gamma(n + 1)$, one sees that the asymptotic relation

$$n! \sim n^n e^{-n}\sqrt{2\pi n}$$

is a special case of the relation, valid for all $|z| \to \infty$:

Γ 8.
$$\Gamma(z) \sim z^{z-1/2}e^{-z}\sqrt{2\pi},$$

uniformly in the sector mentioned above. The twiddle sign $\sim$ means that the quotient of the left-hand side by the right-hand side approaches 1, for $|z| \to \infty$.

Proof of Stirling's Formula

We shall need a simple formula.

Euler Summation Formula. *Let f be any C^1 function of a real variable. Then*

$$\sum_{k=0}^{n} f(k) = \int_0^n f(t)\, dt + \tfrac{1}{2}\left(f(n) + f(0)\right) + \int_0^n P_1(t) f'(t)\, dt.$$

Proof. The sawtooth function looks like this. Note that

$$P_1'(t) = 1$$

for t not an integer.

Graph of $P_1(t)$

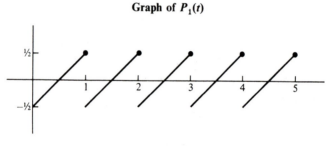

Figure 2

Integrating by parts with $u = P_1(t)$ and $dv = f'(t)\, dt$ yields

$$\int_{k-1}^{k} P_1(t) f'(t)\, dt = P_1(t) f(t) \Big|_{k-1}^{k} - \int_{k-1}^{k} f(t)\, dt$$

$$= \tfrac{1}{2}\left(f(k) + f(k-1)\right) - \int_{k-1}^{k} f(t)\, dt.$$

We take the sum from $k = 1$ to $k = n$. Adding the integral

$$\int_0^n f(t)\, dt \qquad \text{and} \qquad \tfrac{1}{2}\left(f(n) + f(0)\right)$$

then yields the sum $\sum_{k=0}^{n} f(k)$ on the left-hand side and proves the formula.

We now apply the formula to the functions

$$f(t) = \log(z + t), \qquad \text{and} \qquad f(t) = \log(1 + t)$$

and assume until further notice that z is real > 0. Then we have no difficulty dealing with the log and its properties from freshman calculus. Subtracting the expressions in Euler's formula for these two functions, and recalling that

$$\int \log x \, dx = x \log x - x,$$

we obtain

$$\log \frac{z(z+1)\cdots(z+n)}{n!\,(n+1)} = z\log(z+n) + n\log(z+n) - z\log z$$

$$-(z+n) + z + \tfrac{1}{2}(\log(z+n) + \log z)$$

$$-(n+1)\log(n+1) + (n+1)$$

$$-1 + \tfrac{1}{2}\log(n+1)$$

$$+ \text{the terms involving the integrals of } P_1(t).$$

None of this is so bad. We write

$$z\log(z+n) = z\log n\left(1 + \frac{z}{n}\right) = z\log n + z\log\left(1 + \frac{z}{n}\right).$$

The term $z \log n$ is just $\log n^z$, and we move it to the other side.

On the other hand, we note that $n + 1$ occurs in the denominator on the left, and we move $-\log(n+1)$ from the left-hand side to the right-hand side, changing signs. We also make as many cancellations as we can on the right-hand side. We end up with

$$-\left(n + \frac{1}{2}\right)\log\left(1 + \frac{1}{n}\right) + \left(n + \frac{1}{2}\right)\log\left(1 + \frac{z}{n}\right)$$

among other expressions. But from the Taylor expansion for large n (and fixed z) we know that

$$\log\left(1 + \frac{1}{n}\right) = \frac{1}{n} + O\left(\frac{1}{n^2}\right) \qquad \text{and} \qquad \log\left(1 + \frac{z}{n}\right) = \frac{z}{n} + O\left(\frac{1}{n^2}\right).$$

Therefore it is easy to take the limit as n tends to infinity, and we find by **Γ2**,

$$(*) \quad \log \Gamma(z) = \left(z - \frac{1}{2}\right)\log z - z + 1 + \int_0^\infty \frac{P_1(t)}{1+t}\,dt - \int_0^\infty \frac{P_1(t)}{z+t}\,dt.$$

All this is true for positive real z. We shall prove below that the integral on the right is analytic in z for z in the open set U equal to the plane from which the negative real axis has been deleted. Since the other expressions

$$\log \Gamma(z), \qquad z \log z, \qquad \log z, \qquad z$$

are also analytic in this open set (which is simply connected) it follows that formula (∗) is valid for all z in this open set.

Before we deal with the analyticity of the integral on the right, we evaluate the constant

$$1 + \int_0^\infty \frac{P_1(t)}{1 + t}\, dt.$$

We shall prove below that for z pure imaginary, $z = iy$, we have

$$\lim_{y \to \infty} \int_0^\infty \frac{P_1(t)}{iy + t}\, dt = 0.$$

From

$$\Gamma(z)\Gamma(1 - z) = \frac{\pi}{\sin \pi z} \qquad \text{and} \qquad \Gamma(1 - z) = -z\Gamma(-z)$$

we get

$$\Gamma(z)\Gamma(-z) = \frac{-\pi}{z \cdot \sin \pi z}$$

and therefore

$$|\Gamma(iy)| = \sqrt{\frac{2\pi}{y(e^{\pi y} - e^{-\pi y})}}.$$

From (∗) we get

$$1 + \int_0^\infty \frac{P_1(t)}{1 + t}\, dt = \mathrm{Re}\left\{ \log \Gamma(iy) - \left(iy - \frac{1}{2}\right)\log(iy) + iy + \int_0^\infty \frac{P_1(t)}{iy + t}\, dt \right\}$$

$$= \lim_{y \to \infty} \left\{ \log |\Gamma(iy)| + \frac{1}{2}\log y + \frac{\pi y}{2} \right\}$$

$$= \lim_{y \to \infty} \log \sqrt{\frac{2\pi y e^{\pi y}}{y(e^{\pi y} - e^{-\pi y})}}$$

$$= \tfrac{1}{2}\log 2\pi.$$

This proves Stirling's formula, except that we still have to deal with the analyticity of the integral mentioned previously, and the asserted limit. We state these as lemmas.

Let

$$P_2(t) = \tfrac{1}{2}(t^2 - t) \qquad \text{for} \quad 0 \leq t \leq 1,$$

and extend $P_2(t)$ by periodicity to all of $\mathbf{R}$ (period 1). Then $P_2(n) = 0$ for all integers n, and P_2 is bounded. Furthermore, $P_2'(t) = P_1(t)$.

Lemma 2.1. *For z not on the negative real axis, we have*

$$\int_0^\infty \frac{P_1(t)}{z + t}\, dt = \int_0^\infty \frac{P_2(t)}{(z + t)^2}\, dt.$$

The integral is analytic in z in the open set U obtained by deleting the negative real axis from the plane, and one can differentiate under the integral sign on the right in the usual way.

Proof. We write

$$\int_0^\infty = \sum_{n=0}^\infty \int_n^{n+1}.$$

Integrating by parts on each interval $[n, n + 1]$ gives the identity of the lemma. The integral involving P_2 is obviously absolutely convergent, and the differentiation lemma applies.

Observe that differentiating under the integral sign involving P_2 yields a good error term for $\Gamma'/\Gamma(z)$, namely

$$\boxed{\Gamma'/\Gamma(z) = \log z - \frac{1}{z} + 2 \int_0^\infty \frac{P_2(t)}{(z + t)^3}\, dt.}$$

Remark. To integrate by parts more than once, it is more useful to take $P_2(t) = \tfrac{1}{2}(t^2 - t + \tfrac{1}{6})$, the second Bernoulli polynomial, and so forth.

Lemma 2.2.

$$\lim_{y \to \infty} \int_0^\infty \frac{P_1(t)}{iy + t}\, dt = 0.$$

Proof. The limit is clear from Lemma 2.1

EXERCISES XII §2

1. Show that the residue of $\Gamma(z)$ at $-m$ is $(-1)^m/m!$.

2. Prove that for $\mathrm{Re}(z) \geq 0$,

$$\Gamma'/\Gamma(z) = \int_0^\infty \left[\frac{e^{-t}}{t} - \frac{e^{-zt}}{1 - e^{-t}} \right] dt.$$

[*Hint:* Use $1/(z + n) = \int_0^\infty e^{-t(z+n)} \, dt$.]

3. Let C be the contour as shown on Fig. 3. Thus the path consists of $]-\infty, -\epsilon]$, the circle which we denote by K_ϵ, and the path from $-\epsilon$ to $-\infty$. On the plane from which the negative real axis has been deleted, we take the principal value for the log, and for complex s,

$$z^{-s} = e^{-s \log z}.$$

Then

$$\int_C = \int_{-\infty}^{-\epsilon} + \int_{K_\epsilon} + \int_{-\epsilon}^{-\infty}.$$

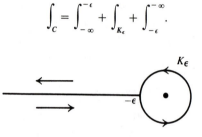

Figure 3

The integrals will involve z^s, and the two values for z^s in the first and third integral will differ by a constant.

(a) Prove that the integral

$$\int_C e^{-z} z^{-s} \, dz$$

defines an entire function of s.

(b) Prove that for $\mathrm{Re}(1 - s) > 0$ we have

$$\int_C e^{-z} z^{-s} \, dz = 2i \sin \pi s \int_0^\infty e^{-u} u^{-s} \, du.$$

(c) Show that

$$\frac{1}{\Gamma(s)} = \frac{1}{2\pi i} \int_C e^z z^{-s} \, dz.$$

The contour integral gives another analytic continuation for $1/\Gamma(s)$ to the whole plane.

4. Let

$$G(z) = \frac{e^{-z}}{1 - e^{-z}} \quad \text{and} \quad F(z) = \frac{ze^z}{e^z - 1},$$

so that

$$G(-z) = -\frac{1}{z} F(z).$$

Let

$$H(s) = \int_C \frac{1}{z} F(z) z^s \frac{dz}{z},$$

where the contour C is the same as in Exercise 3.
(a) Prove that H is an entire function.
(b) Changing variable, putting $z = -w$, and letting $\epsilon \to 0$ show that

$$H(s) = -(e^{\pi is} - e^{-\pi is}) \int_0^\infty G(t) t^s \frac{dt}{t}.$$

This is called **Hankel's integral**.

5. Define $\zeta(s) = \sum 1/n^s$ for $\text{Re}(s) > 1$.
(a) Show that

$$\Gamma(s)\zeta(s) = \int_0^\infty G(t) t^s \frac{dt}{t}.$$

The expression in Exercise 4 gives the analytic continuation of $\zeta(s)$ to the whole plane, as a meromorphic function.
(b) Show that $\zeta(s)$ is holomorphic except for a simple pole at $s = 1$.
(c) Show that the residue of $\zeta(s)$ at $s = 1$ is 1.
(d) Show that $\zeta(s)$ has zeros of order 1 at the negative integers.

6. **Mellin Inversion Formula.** Show that for $x > 0$ we have

$$e^{-x} = \frac{1}{2\pi i} \int_{\sigma = \sigma_0} x^{-s} \Gamma(s)\, ds,$$

where $s = \sigma + it$, and the integral is taken on a vertical line with fixed real part σ_0, and $-\infty < t < \infty$. [*Hint:* What is the residue of $x^{-s}\Gamma(s)$ at $x = -n$?]

Let the **Paley–Wiener space** consist of those entire functions f for which there exists a positive number C having the following property. Given an integer $N > 0$, we have

$$|f(x + iy)| \ll \frac{C^{|x|}}{(1 + |y|)^N},$$

where the implied constant in $\ll$ depends on f and N. We may say that f is at most of exponential growth with respect to x, and is rapidly decreasing, uniformly in every vertical strip of finite width.

7. If f is C^∞ (infinitely differentiable) on the open interval $]0, \infty[$, and has compact support, then its **Mellin transform** Mf defined by

$$Mf(z) = \int_0^\infty f(t)t^z \frac{dt}{t}$$

is in the Paley–Wiener space. [*Hint*: Integrate by parts.]

8. Let F be in the Paley–Wiener space. For any real x, define the function $'M_x F$ by

$$'M_x F(t) = \int_{\text{Re}\,z = x} F(z)t^z \frac{dz}{i}.$$

The integral is supposed to be taken on the vertical line $z = x + iy$, with fixed x, and $-\infty < y < \infty$. Show that $'M_x F$ is independent of x, so can be written $'MF$. [*Hint*: Use Cauchy's theorem.] Prove that $'M_x F$ has compact support on $]0, \infty[$.

Remark. If you want to see these exercises worked out, cf. $SL_2(\mathbf{R})$, Chapter V, §3. The two maps M and $'M$ are inverse to each other, but one needs the Fourier inversion formula to prove this.

9. Let a, b be real numbers > 0. Define

K 1.
$$K_s(a, b) = \int_0^\infty e^{-(a^2 t + b^2/t)} t^s \frac{dt}{t}.$$

Prove that K_s is an entire function of s. Prove that

K 2.
$$K_s(a, b) = (b/a)^s K_s(ab),$$

where for $c > 0$ we define

K 3.
$$K_s(c) = \int_0^\infty e^{-c(t + 1/t)} t^s \frac{dt}{t}.$$

Prove that

K 4.
$$K_s(c) = K_{-s}(c).$$

K 5.
$$K_{1/2}(c) = \sqrt{\pi/c}\, e^{-2c}.$$

[*Hint*: Differentiate the integral for $\sqrt{x}\, K_{1/2}(x)$ under the integral sign.] Let $x_0 > 0$ and $\sigma_0 \leq \sigma \leq \sigma_1$. Show that there exists a number $C = C(x_0, \sigma_0, \sigma_1)$ such that if $x \geq x_0$, then

K 6.
$$K_\sigma(x) \leq Ce^{-2x}.$$

Prove that

K 7.
$$\int_{-\infty}^{\infty} \frac{1}{(u^2 + 1)^s} \, du = \sqrt{\pi} \, \frac{\Gamma(s - 1/2)}{\Gamma(s)}$$

for $\text{Re}(s) > 1/2$. Also prove that

K 8.
$$\Gamma(s) \int_{-\infty}^{\infty} \frac{e^{ixu}}{(u^2 + 1)^s} \, du = 2\sqrt{\pi}(x/2)^{s - 1/2} K_{s - 1/2}(x)$$

for $\text{Re}(s) > 1/2$.
[To see this worked out, cf. my *Elliptic Functions*, Chapter 20.]

CHAPTER XIII

Analytic Continuation

In this chapter we give further means to extend the domain of definition of an analytic function. We shall apply Theorem 1.2 of Chapter III in the following context. Suppose we are given an analytic function f on an open connected set U. Let V be open and connected, and suppose that $U \cap V$ is not empty, so is open. We ask whether there exists an analytic function g on V such that $f = g$ on $U \cap V$, or only such that $f(z) = g(z)$ for all z in some set of points of $U \cap V$ which is not discrete. The above-mentioned Theorem 1.2 shows that such a function g if it exists is uniquely determined. One calls such a function g a **direct analytic continuation** of f, and we also say that (g, V) is a **direct analytic continuation** of (f, U). We use the word "direct" because later we shall deal with analytic continuation along a curve and it is useful to have an adjective to distinguish the two notions. For simplicity, however, one usually omits the word "direct" if no confusion can result from this omission. If a direct analytic continuation exists as above, then there is a unique analytic function h on $U \cup V$ such that $h = f$ on U and $h = g$ on V.

In the first example, Schwarz symmetry, we even generalize analytic continuation to the case when V and U are disjoint, but have a curve in common at their boundary.

XIII §1. Schwarz Reflection

Theorem 1.1. *Let U^+ be an open set in the upper half plane, and suppose that the boundary of U^+ contains an open interval I of real numbers. Let U^- be the reflection of U^+ across the real axis (i.e. the*

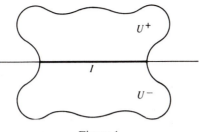

Figure 1

set of $\bar{z}$ with $z \in U^+$), and as in Figure 1, let

$$U = U^+ \cup I \cup U^-.$$

(i)　*If f is a function on U, analytic on U^+ and U^-, and continuous on I, then f is analytic on U.*

(ii)　*If f is a function on $U^+ \cup I$, analytic on U^+ and continuous on I, and f is real valued on I, then f has a unique analytic continuation F on U, and F satisfies*

$$F(z) = \overline{f(\bar{z})}.$$

Proof. We reduce (ii) to (i) if we define F by the above formula. From the hypothesis that f is real valued on I, it is clear that F is continuous at all points of I, whence on U. Furthermore, from the power series expansion of f at some point z_0 in the upper half plane, it is immediate from the formula that F is analytic on U^-. There remains to prove (i).

We consider values of z near I, and especially near some point of I. Such values lie inside a rectangle, as shown on Fig. 2(a). This rectangle has a boundary $C = C^+ + C^-$, oriented as shown. We define

$$g(z) = \frac{1}{2\pi i} \int_C \frac{f(\zeta)}{\zeta - z}\, d\zeta,$$

for z inside the rectangle. Then for z not on I, we have

$$g(z) = \frac{1}{2\pi i} \int_{C^+} \frac{f(\zeta)}{\zeta - z}\, d\zeta + \frac{1}{2\pi i} \int_{C^-} \frac{f(\zeta)}{\zeta - z}\, d\zeta.$$

Let C_ϵ^+ be the rectangle as shown on Fig. 2(b). Then for z inside C_ϵ^+, Cauchy's formula gives

$$f(z) = \frac{1}{2\pi i} \int_{C_\epsilon^+} \frac{f(\zeta)}{\zeta - z}\, d\zeta.$$

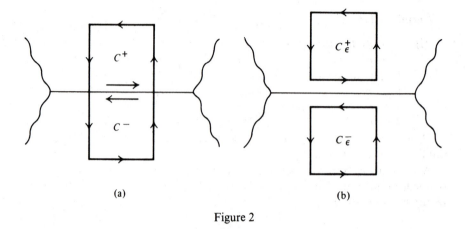

Figure 2

An easy continuity argument shows that taking the limit as $\epsilon \to 0$, we get in fact

$$f(z) = \frac{1}{2\pi i} \int_{C^+} \frac{f(\zeta)}{\zeta - z} \, d\zeta,$$

for z inside C_ϵ^+, and hence inside C^+. On the other hand, a similar argument combined with Cauchy's theorem shows that

$$\int_{C^-} \frac{f(\zeta)}{\zeta - z} \, d\zeta = 0.$$

Hence $g(z) = f(z)$ if z is in U near I. By continuity, we also obtain that $g(z) = f(z)$ if z is on I. By symmetry, the same arguments would show that $g(z) = f(z)$ if z is in U^-, and z is near I. This proves that $g(z) = f(z)$ for z near I in U, and hence that g is the analytic continuation of f in U, as was to be shown.

The theorem applies to more general situations which are analytically isomorphic to that of the theorem. More precisely, let V be an open set in the complex numbers, and suppose that V is the disjoint union

$$V = V^+ \cup \gamma \cup V^-,$$

where V^+ is open, V^- is open, and γ is some curve. We suppose that there exists an analytic isomorphism

$$\psi : U \to V$$

such that

$$\psi(U^+) = V^+, \qquad \psi(I) = \gamma, \qquad \psi(U^-) = V^-.$$

Corollary 1.2.

(i)   *Given a function g on V which is analytic on V^+ and V^-, and continuous on γ, then g is analytic on V.*

(ii) *If g is an analytic function on V^+ which extends to a continuous function on $V^+ \cup \gamma$, and is real valued on γ, then g extends to an analytic function on V.*

Proof. Obvious, using successively parts (i) and (ii) of the theorem, applied to the function $f = g \circ \psi$.

A standard example of this situation occurs when V^+ is an open set on one side of an arc of the unit circle, as on Fig. 3(b). It is then useful to remember that the map

$$z \mapsto 1/z$$

interchanges the inside and outside of the unit circle.

We can also perform an analytic continuation in the more general situation, as in part (ii) of the theorem. More generally, let us define a curve

$$\gamma: [a, b] \to \mathbf{C}$$

to be **real analytic** if for each point t_0 in $[a, b]$ there exists a convergent power series expansion

$$\gamma(t) = \sum a_n(t - t_0)^n$$

for all t sufficiently close to t_0. Using these power series, we see that γ extends to an analytic map of some open neighborhood of $[a, b]$. We

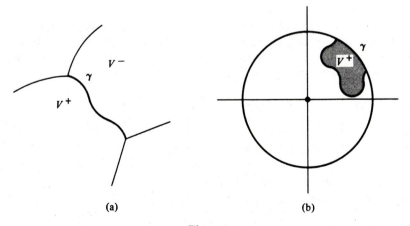

(a)   (b)

Figure 3

shall say that γ is a **proper analytic arc** if γ is injective, and if $\gamma'(t) \neq 0$ for all $t \in [a, b]$. We leave it as an exercise to prove:

If γ is a proper analytic arc, then there exists an open neighborhood W of $[a, b]$ such that γ extends to an analytic isomorphism of W.

Let $\gamma: [a, b] \to \mathbf{C}$ be a proper analytic arc, which is contained in the boundary of an open set U (in other words, the image of γ is contained in the boundary of U). We shall say that U **lies on one side of** γ if there exists an extension γ_W of γ to some open neighborhood W of $[a, b]$ as above, such that $\gamma_W^{-1}(U)$ lies either in the upper half plane, or in the lower half plane. As usual, $\gamma_W^{-1}(U)$ is the set of points $z \in W$ such that $\gamma_W(z) \in U$.

Let f be an analytic function on an open set U. Let γ be an analytic arc which is contained in the boundary of U, and such that U lies on one side of γ. We say that f has an **analytic continuation across** γ if there exists an open neighborhood W of γ (without its end points) such that f has an analytic continuation to $U \cup W$.

Theorem 1.3. *Let f be analytic on an open set U. Let γ be a proper analytic arc which is contained in the boundary of U, and such that U lies on one side of γ. Assume that f extends to a continuous function on $U \cup \gamma$ (i.e. $U \cup \text{Image } \gamma$), and that $f(\gamma)$ is also an analytic arc, such that $f(U)$ lies on one side of $f(\gamma)$. Then f has an analytic continuation across γ.*

Proof. There exist analytic isomorphisms

$$\varphi: W_1 \to \text{neighborhood of } \gamma,$$

$$\psi: W_2 \to \text{neighborhood of } f(\gamma),$$

where W_1, W_2 are open sets as illustrated on Fig. 4, neighborhoods of real intervals I_1 and I_2, respectively. These open sets are selected sufficiently small that U and $f(U)$ lie on one side of W_1, W_2, respectively.

Define

$$g = \psi^{-1} \circ f \circ \varphi.$$

Then g is defined and analytic on one side of the interval I_1, and is also continuous on I_1 (without its end points). Furthermore g is real valued. Hence g has an analytic continuation by Theorem 1.1 to the other side of I_1. The function

$$\psi \circ g \circ \varphi^{-1}$$

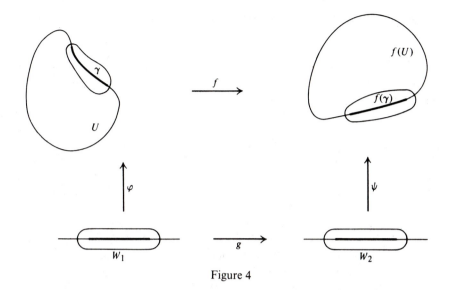

Figure 4

then defines an analytic continuation of f on some open neighborhood of γ (without its end points), as desired.

The analytic continuation of f in Theorem 1.3 is also often called the **reflection** of f across γ.

EXERCISES XIII §1

1. Let C be an arc of unit circle $|z| = 1$, and let U be an open set inside the circle, having that arc as a piece of its boundary. If f is analytic on U and continuous on C, and takes real values on C, show that f can be continued across C by the relation

$$f(z) = \overline{f(1/\bar{z})}.$$

2. Suppose, on the other hand, that instead of taking real values on C, f takes on values on the unit circle, that is,

$$|f(z)| = 1 \qquad \text{for} \quad z \text{ on } C.$$

Show that the analytic continuation of f across C is now given by

$$f(z) = 1/\overline{f(1/\bar{z})}.$$

3. Let f be a function which is continuous on the closed unit disc and analytic on the open disc. Assume that $|f(z)| = 1$ whenever $|z| = 1$. Show that the function f can be extended to a meromorphic function, with at most a finite number of poles in the whole plane.

XIII §2. Continuation Along a Path

Let f be analytic on a disc D_0 centered at a point z_0. Let γ be a path whose beginning point is z_0 and whose end point is w, and say γ is defined on the interval $[a, b]$. In this section, smoothness of the path plays no role, and "path" will mean continuous path. Let

$$a = a_0 \leq a_1 \leq a_2 \leq \cdots \leq a_n = b$$

be a partition of the interval. Let D_i be a disc containing $\gamma(a_i)$ as shown on Fig. 5. Instead of discs, we could also take convex open sets. The intersection of two convex sets is also convex. Actually, what we shall need precisely is that the intersection of a finite number of the sets $D_0, \ldots, D_n$ is connected if it is not empty.

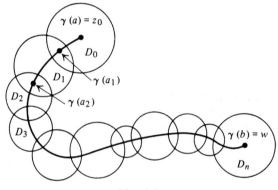

$\gamma(a) = z_0$

D_0

D_1 $\gamma(a_1)$

D_2 $\gamma(a_2)$

D_3

$\gamma(b) = w$

D_n

Figure 5

We shall say that the sequence

$$\{D_0, D_1, \ldots, D_n\}$$

is **connected by the curve along the partition** if the image $\gamma([a_i, a_{i+1}])$ is contained in D_i. Then the intersection $D_i \cap D_{i+1}$ contains $\gamma(a_{i+1})$.

Let f_0 be analytic on D_0. By an **analytic continuation of** (f_0, D_0) **along a connected sequence** $[D_0, \ldots, D_n]$ we shall mean a sequence of pairs

$$(f_0, D_0), (f_1, D_1), \ldots, (f_n, D_n)$$

such that (f_{i+1}, D_{i+1}) is a direct analytic continuation of (f_i, D_i) for $i = 0, \ldots, n$. This definition appears to depend on the choice of partition and the choice of the connected sequence $\{D_0, \ldots, D_n\}$. We shall prove:

Theorem 2.1. *Let* $(g_0, E_0), \ldots, (g_m, E_m)$ *be an analytic continuation of* (g_0, E_0) *along a connected sequence* $\{E_0, \ldots, E_m\}$ *with respect to a* *partition of the path* γ. *If* $f_0 = g_0$ *in some neighborhood of* z_0, *then* $g_m = f_n$ *in some neighborhood of* $\gamma(b)$, *so* (g_m, E_m) *is a direct analytic continuation of* (f_n, D_n).

Thus we obtain a well-defined analytic function in a neighborhood of the end point of the path, which is called the (f_0, D_0) **along the path** γ. As a matter of notation, we may also denote this function by f_γ.

We now give the proof that the analytic continuation does not depend on the choices of partition and discs, in a way similar to that used in Chapter III, §4 when we dealt with the integral along a path. The proof here is equally easy and straightforward.

Suppose first that the partition is fixed, and let

$$(g_0, E_0), \ldots, (g_n, E_n)$$

be an analytic continuation along another connected sequence

$$\{E_0, \ldots, E_n\}.$$

Suppose $g_0 = f_0$ in a neighborhood of z_0, which means that (g_0, E_0) is a direct analytic continuation of (f_0, D_0). Since $D_0 \cap E_0$ is connected it follows that $f_0 = g_0$ on the whole set $D_0 \cap E_0$, which also contains z_1. By hypothesis, $f_1 = f_0$ on $D_0 \cap D_1$ and $g_1 = g_0$ on $E_0 \cap E_1$. Hence $f_1 = g_1$ on $D_0 \cap E_0 \cap D_1 \cap E_1$ which contains z_1. Since $D_1 \cap E_1$ is connected, it follows that $f_1 = g_1$ on $D_1 \cap E_1$. Thus (g_1, E_1) is a direct analytic continuation of (f_1, D_1). We can now proceed by induction to see that (g_n, E_n) is a direct analytic continuation of (f_n, D_n), thus concluding the proof in this case.

Next let us consider a change in the partition. Any two partitions have a common refinement. To show the independence of the partition it suffices to do so when we insert one point in the given partition, say we insert c in the interval $[a_k, a_{k+1}]$ for some k. On one hand, we take the connected sequence

$$\{D_0, \ldots, D_k, D_k, \ldots, D_n\},$$

where D_k is repeated twice, so that $\gamma([a_k, c]) \subset D_k$ and $\gamma([c, a_{k+1}]) \subset D_k$. Then

$$(f_0, D_0), \ldots, (f_k, D_k), (f_k, D_k), \ldots, (f_n, D_n)$$

is an analytic continuation of (f_0, D_0) along this connected sequence. On the other hand, suppose that

$$(g_0, E_0), \ldots, (g_k, E_k), (g_k^*, E_k^*), \ldots, (g_n, E_n)$$

is an analytic continuation of (g_0, E_0) along another connected sequence $\{E_0, \ldots, E_k, E_k^*, \ldots, E_n\}$ with respect to the new partition, and $g_0 = f_0$ in some neighborhood of z_0. By the first part of the proof, we know that (g_k, E_k) is a direct analytic continuation of (f_k, D_k). By hypothesis, g_k^* is equal to g_k on $E_k \cap E_k^*$, $f_k = g_k$ on $D_k \cap E_k$, so

$$g_k^* = f_k \qquad \text{on} \quad D_k \cap E_k \cap E_k^*, \qquad \text{which contains} \quad z_{k+1} = \gamma(a_{k+1}).$$

Therefore (g_k^*, E_k^*) is a direct analytic continuation of (f_k, D_k). Again we can apply the first part of the proof to the second piece of the path which is defined on the interval $[a_{k+1}, a_n]$, with respect to the partition $[a_{k+1}, a_{k+2}, \ldots, a_n]$ to conclude the proof of the theorem.

Example. Let us start with the function $\log z$ defined by the usual power series on the disc D_0 which is centered at 1 and has radius < 1 but > 0. Let the path be the circle of radius 1 oriented counterclockwise as usual. If we continue $\log z$ along this path, and let (g, D) be its continuation, then near the point 1 it is easy to show that

$$g(z) = \log z + 2\pi i.$$

Thus g differs from f_0 by a constant, and is not equal to f_0 near $z_0 = 1$.

Example. Let

$$f(z) = e^{(1/2)\log z},$$

where f is defined in a neighborhood of 1 by the principal value for the log. [We could write $f(z) = \sqrt{z}$ in a loose way, but the square root sign has the usual indeterminacy, so it is meaningless to use the expression $\sqrt{z}$ for a function unless it is defined more precisely.] We may take the analytic continuation of f along the unit circle. The analytic continuation of $\log z$ along the circle has the value

$$\log z + 2\pi i$$

near 1. Hence the analytic continuation of f along this circle has the value

$$g(z) = e^{(1/2)(\log z + 2\pi i)} = -e^{(1/2)\log z} = -f(z)$$

for z near 1. This is the other solution to the equation

$$f(z)^2 = z$$

near 1. If we continue g analytically once more around the circle, then we obtain $f(z)$, the original function.

Remark. In most texts, one finds a picture of the "Riemann surface" on which "$\sqrt{z}$" is defined, and that picture represents two sheets crossing themselves, and looking as if there was some sort of singularity at the origin. **It should be emphasized that the picture is totally and irretrievably misleading.** The proper model for the domain of definition of $f(z)$ is obtained by introducing another plane, so that the correspondence between $f(z)$ and z is represented by the map

$$\mathbf{C} \to \mathbf{C}$$

given by $\zeta \mapsto \zeta^2$. The function φ such that $\varphi(\zeta) = \zeta$ is perfectly well defined on $\mathbf{C}$. The association $\zeta \mapsto \zeta^2$ gives a double covering of $\mathbf{C}$ by $\mathbf{C}$ at all points except the origin (it is the function already discussed in Chapter I, §3).

More generally, let P be a polynomial in two variables, $P = P(T_1, T_2)$, not identically zero. A solution (analytic) $f(z)$ of the equation

$$P(f(z), z) = 0$$

for z in some open set is called an **algebraic function**. It can be shown that if we delete a finite number of points from the plane, then such a solution f can be continued analytically along every path. Furthermore, we have the following theorem.

Theorem 2.2. *Let $P(T_1, T_2)$ be a polynomial in two variables. Let γ be a curve with beginning point z_0 and end point w. Let f be analytic at z_0, and suppose that f has an analytic continuation f_γ along the curve γ. If*

$$P(f(z), z) = 0 \quad \text{for} \quad z \text{ near } z_0,$$

then

$$P(f_\gamma(z), z) = 0 \quad \text{for} \quad z \text{ near } w.$$

Proof. This is obvious, because the relation holds in each successive disc $D_0, D_1, \ldots, D_m$ used to carry out the analytic continuation.

Theorem 2.3 (Monodromy Theorem). *Let U be a connected open set. Let f be analytic at a point z_0 of U, and let γ, η be two paths from z_0 to a point w of U. Assume:*

(i) *γ is homotopic to η in U.*
(ii) *f can be continued analytically along any path in U.*

Let f_γ, f_η be the analytic continuations of f along γ and η, respectively. Then f_γ and f_η are equal in some neighborhood of w.

Proof. Let η^- be the reverse path to η. Then $\{\gamma, \eta^-\}$ is a closed path. Let

$$\psi \colon [a, b] \times [c, d] \to U$$

be the homotopy, whose domain of definition is the rectangle $R^{(0)}$ illustrated on Fig. 6. The images of the vertical segments under ψ are equal to z_0 and w, respectively. The images of the top and bottom segment under ψ are equal to γ and η^-, respectively (the orientation being as shown). We assume that $f_\gamma \neq f_\eta$, and derive a contradiction.

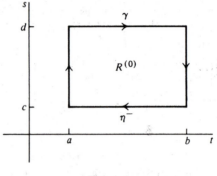

Figure 6

We divide the rectangle into four congruent rectangles R_1, R_2, R_3, R_4 as shown on Fig. 7, giving rise to four closed paths as shown (actually, the images under ψ). (The paths in U are actually the images under ψ of those shown in the diagram.) Each new closed path consists of some line segments occurring with one orientation, and also with the reverse orientation, and the boundary of a rectangle (oriented clockwise). Since $f_\gamma \neq f_\eta$, there exists at least one rectangle, say $R^{(1)}$, such that the continuation of f along the path image under ψ of

$$\{L_1, \partial R^{(1)}, L_1^-\}$$

is not equal to f. Here L_1 denotes the sequence of line segments leading from the lower left corner to the lower left corner of the rectangle $R^{(1)}$.

In this manner, by performing the construction repeatedly, we obtain a sequence of rectangles

$$R^{(0)} \supset R^{(1)} \supset R^{(2)}$$

and a path of line segments L_n from (a, c) to the lower left corner of $R^{(n)}$, giving rise to the closed path

$$\sigma_n = \{L_n, \partial R^{(n)}, L_n^-\},$$

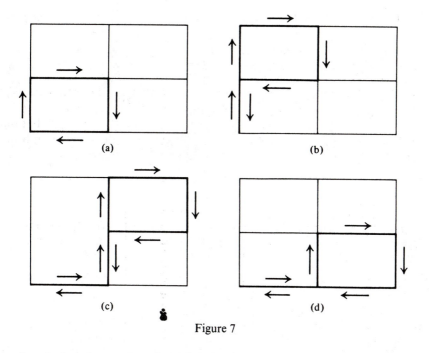

Figure 7

and such that the continuation of f along $\psi(\sigma_n)$ is not equal to f. Observe that L_n is part of L_{n+1}.

The intersection of all the rectangles $\cap R^{(n)}$ is a point (s^*, t^*), and the paths L_n approach a limiting path L, which is a continuous path from (a, c) to (s^*, t^*) as in Fig. 8. We have $\psi(a, c) = z_0$ and we let

$$\psi(s^*, t^*) = z^*.$$

The image $\psi(L) = \sigma$ is a continuous path from $\psi(a, c) = z_0$ to the point

$$\psi(s^*, t^*) = z^*$$

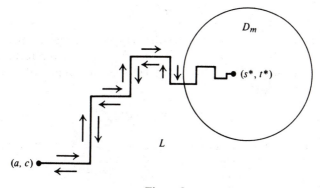

Figure 8

in U. By hypothesis f can be continued along $\psi(\sigma)$. Let $D_0,\ldots,D_m$ be a sequence of discs connected along a partition of $\psi(\sigma)$, such that f has analytic continuation f_i on D_i, and so analytic continuation $f_{\psi(\sigma)}$ along $\psi(\sigma)$. For all n sufficiently large, L consists of L_n and a small path lying entirely inside D_m. Furthermore, taking n sufficiently large, the rectangle $R^{(n)}$ is also contained inside D_m. Thus the analytic continuation of f along $\psi(\sigma)$ consists of the analytic continuation of f along $\psi(L_n)$, and one further continuation to the final disc D_m. Having continued f along $\psi(L_n)$, we conclude that the further continuation along the image of the boundary of $R^{(n)}$ is still equal to $f_{\psi(L_n)}$. Making the further continuation along $\psi(L_n^-)$ we see that $f_{\psi(\sigma_n)} = f$, a contradiction which proves the theorem.

Example. Let g be an analytic function on a connected open set U. We do not assume that U is simply connected. Let $z_0 \in U$. Let f_0 be a primitive of g in some disc containing z_0, so

$$f_0(z) = \int_{z_0}^{z} g(\zeta)\, d\zeta,$$

where the integral is taken along any path from z_0 to z inside the disc. Then f_0 can be analytically continued along any path in U, essentially by integration along the path. For instance let γ be a path, and let z_1 be a point on γ. Say

$$\gamma: [a, b] \to U, \qquad \text{with} \quad \gamma(a) = z_0.$$

and $z_1 = \gamma(c)$ with $a < c \leq b$. Then we can define f_1 in a neighborhood of z_1 by taking the integral of f along the path from z_0 to z_1, that is along the restriction of γ to the interval $[a, c]$, and then from z_1 to z along any path contained in a small disc D containing z_1, as illustrated in Fig. 9.

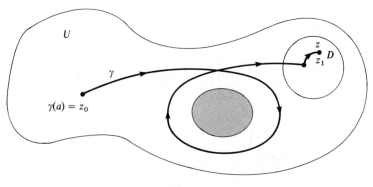

Figure 9

In Fig. 9, we have drawn U with a shaded hole, so U is not simply connected. The monodromy theorem applies to this situation.

A concrete case can be given as follows. Let $R(z)$ be a rational function, that is a quotient of polynomials. We can factor the numerator and denominator into linear factors, say

$$R(z) = \frac{(z - a_1) \cdots (z - a_n)}{(z - b_1) \cdots (z - b_m)}.$$

Then R has poles at $b_1, \ldots, b_m$ (which we do not assume all distinct). Let U be the plane from which the numbers $b_1, \ldots, b_m$ are deleted. Then U is connected but not simply connected if $m \geq 1$. We can define the integral

$$f(z) = \int_{z_0, \gamma}^{z} R(\zeta)\, d\zeta,$$

along a path from z_0 to z. This integral depends on the path, and in fact depends on the winding numbers of the path around the points $b_1, \ldots, b_m$.

Example. The preceding example can be generalized, we do not really need g to be analytic on U; we need only something weaker. For instance, let U be the plane from which the three points 1, 2, 3 have been deleted. Then for any simply connected open set V in U, there is a function

$$[(z - 1)(z - 2)(z - 3)]^{1/2},$$

i.e. a square root of $(z - 1)(z - 2)(z - 3)$, and any two such square roots on V differ by an additive constant. Starting from a point z_0 in U, let

$$g_0(z) = \frac{1}{[(z - 1)(z - 2)(z - 3)]^{1/2}},$$

where the square root is one of the possible determinations analytic in a neighborhood of z_0. Then we can form the integral

$$\int_{z_0}^{z} g_0(\zeta) d\zeta,$$

along any path in U from z_0 to z. The integral will depend on the path, but the mondromy theorem applies to this situation as well.

EXERCISES XIII §2

1. Let f be analytic in the neighborhood of a point z_0. Let k be a positive integer, and let $P(T_1, \ldots, T_k)$ be a polynomial in k variables. Assume that

$$P(f, Df, \ldots, D^k f) = 0,$$

where $D = d/dz$. If f can be continued along a path γ, show that

$$P(f_\gamma, Df_\gamma, \ldots, D^k f_\gamma) = 0.$$

2. **(Weierstrass).** Prove that the function

$$f(z) = \sum z^{n!}$$

cannot be analytically continued to any open set strictly larger than the unit disc. [*Hint*: If z tends to 1 on the real axis, the series clearly becomes infinite. Rotate z by a k-th root of unity for positive integers k to see that the function becomes infinite on a dense set of points on the unit circle.]

3. Give another proof for the monodromy theorem, based on the concept of curves which are close together, as in Chapter III, §5.

4. **The Dilogarithm.** Let

$$f(z) = \frac{\log(1 - z)}{z} = \sum_{n=1}^{\infty} \frac{z^{n-1}}{n}$$

in a neighborhood of the origin, and otherwise the function is defined by analytic continuation on the simply connected set obtained by deleting the set of real

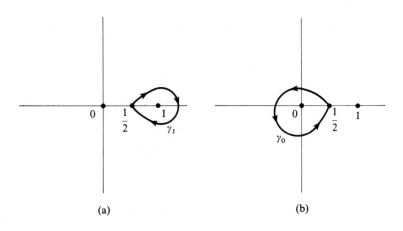

(a)   (b) numbers $x \geq 1$ from the plane. Then f is analytic on this open set. Let $z_0 = \frac{1}{2}$, for instance. Define the **dilogarithm**

$$L_2(z) = \int_{z_0}^{z} f(\zeta) \, d\zeta,$$

taken over any path in this open set.

(a) Now let γ_1 be a curve as shown on Fig. 10(a), circling 1 exactly once. How does the analytic continuation of L_2 along γ_1 differ from L_2 in a neighborhood of z_0?

(b) How does the analytic continuation of L_2 along the path γ_0 on Fig. 10(b) differ from L_2?

(c) If you continue L_2 first around γ_0 and then around γ_1, how does this continuation differ from continuing L_2 first around γ_1 and then around γ_0? [They won't be equal!]

CHAPTER XIV

The Riemann Mapping Theorem

In this chapter we give the general proof of the Riemann mapping theorem, and also state results concerning the behavior at the boundary.

XIV §1. Statement and Application to Picard's Theorem

The **Riemann mapping theorem** asserts:

Let U be a simply connected open set which is not the whole plane. Then U is analytically isomorphic to the disc of radius 1. More precisely, given $z_0 \in U$, there exists an analytic isomorphism

$$f: U \to D(0, 1)$$

of U with the unit disc, such that $f(z_0) = 0$. Such an isomorphism is uniquely determined up to a rotation, i.e. multiplication by $e^{i\theta}$ for some real θ, and is therefore uniquely determined by the additional condition

$$f'(z_0) > 0.$$

We have seen in Chapter VII, §2 that the only analytic automorphisms of the disc are given by the mappings

$$w \mapsto e^{i\varphi} \frac{w - \alpha}{1 - \bar{\alpha}w},$$

where $|\alpha| < 1$ and φ is real. This makes the uniqueness statement obvious: If f, g are two analytic isomorphisms of U and $D(0, 1)$ satisfying the prescribed condition, then $f \circ g^{-1}$ is an analytic automorphism of the disc leaving the origin fixed and having positive derivative at the origin. It is then clear from the above formula that $f \circ g^{-1} = \text{id}$.

Before giving the existence proof, we give an application to Picard's theorem. We begin with a relevant example of an analytic isomorphism involving circles. Let C_1, C_2 be two circles which intersect in two points z_1, z_2 and are perpendicular to each other at these points, as shown on Fig. 1. We suppose that C_2 does not go through the center of C_1. Let T be inversion through C_1. By Theorem 5.2, Chapter VII, T maps C_2 on another circle, and preserves orthogonality. Since every point of C_1 is fixed, and since given two points on C_1 there is exactly one circle passing through these two points and perpendicular to C_1, it follows that T maps C_2 onto itself. A point α on C_2 which lies outside C_1 is mapped on a point α' again on C_2 but inside C_1. Thus T interchanges the two arcs of C_2 lying outside and inside C_1 respectively.

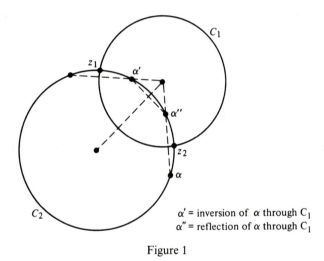

$\alpha' = $ inversion of α through C_1
$\alpha'' = $ reflection of α through C_1

Figure 1

For the next example, it is convenient to make a definition. Let U be the open set bounded by three circular arcs perpendicular to the unit circle, as on Fig. 2. We shall call U a **triangle**. We suppose that the circular sides of the triangle do not pass through the center.

The remark implies that the reflection of U across any one of its "sides" is again a triangle, whose sides are circular arcs perpendicular to the unit circle.

Take three equidistant points on the unit circle, and join them with circular arcs perpendicular to the unit circle at these points. The region bounded by these three arcs is what we have called a triangle U, as shown on Fig. 3. If we invert U across any one of its sides, we obtain another

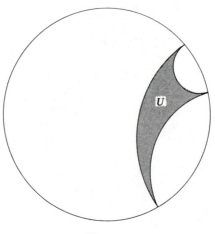

Figure 2

triangle U'. Inverting successively such triangles, we obtain a figure such as is illustrated in Fig. 4.

Let

$$g: U \to H \qquad \text{and} \qquad h: H \to U$$

be inverse analytic isomorphisms between the triangle U and the upper half plane. Such isomorphisms can always be found by the Riemann mapping theorem. Furthermore, we can always select h such that the three vertices z_0, z_1, z_∞ are mapped on $0, 1, \infty$, respectively (why?). We shall prove in §4 that g can be extended continuously to the boundary

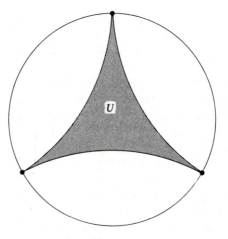

Figure 3

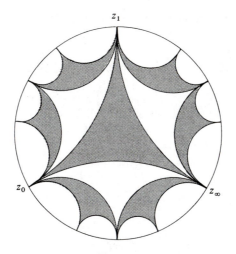

Figure 4

of U, which consists of three very smooth arcs. It follows that g maps the three arcs on the intervals

$$[-\infty, 0], \qquad [0, 1], \qquad [1, \infty]$$

respectively, and in particular, g is real valued on each arc.

By the Schwarz reflection principle, one may continue g analytically across each arc, or vice versa, one may continue h analytically across each interval, by reflection across the sides of the triangle. If h_γ denotes the analytic continuation of h along any path γ not passing through 0, 1, then for any z on the real line, $z \neq 0$, 1 we have $h_\gamma(z)$ on the side of some iterated reflection of the original triangle. For any complex number $z \neq 0$, 1 the value $h_\gamma(z)$ lies inside the unit disc D.

Picard's Theorem. *Let f be an entire function whose values omit at least two complex numbers. Then f is constant.*

Proof. After composing f with a fractional linear map, we may assume that the omitted values are 0, 1. Without loss of generality, we may assume that there is some z_0 such that $f(z_0)$ lies in the upper half plane. (Otherwise, we would proceed in a similar manner relative to the lower half plane.) The analytic function $h(f(z))$ for z near z_0 maps a neighborhood of z_0 inside U. We may continue this function $h \circ f$ analytically along any path in $\mathbf{C}$, because f may be so continued, and the image of any path in $\mathbf{C}$ under h is a path which does not contain 0 or 1. Since $\mathbf{C}$ is simply connected, the analytic continuation of $h \circ f$ to $\mathbf{C}$ is then well defined on all of $\mathbf{C}$, and its values lie in the unit disc, so are bounded. By

Liouville's theorem, we conclude that $h \circ f$ is constant, whence f is constant, which proves the theorem.

Remark. The crucial step in the proof was the use of the functions g and h. In classical times, they did not like functions defined essentially in an "abstract nonsense" manner. The situation of the triangle is so concrete that one would prefer to exhibit the desired functions explicitly. This can be done, and was done in Picard's original proof. He used the standard "modular function" denoted classically by λ. Its properties and definition follow easily from the theory of elliptic functions. Cf. for instance my book on *Elliptic Functions*, Chapter 18, §5. It is usually easier to deal with the upper half plane rather than the unit disc in such concrete situations, because the upper half plane is the natural domain of definition of the classical modular functions.

XIV §2. Compact Sets in Function Spaces

Let U be an open set. We denote by $\mathrm{Hol}(U)$ the space of homomorphic functions on U. A subset Φ of $\mathrm{Hol}(U)$ will be called **relatively compact** if every sequence in Φ has a subsequence which converges uniformly on every compact subset of U, not necessarily to an element of Φ itself. (*Note*: Instead of relatively compact, one sometimes calls such subsets **normal**, or **normal families**. The word relatively compact fits general notions of metric spaces somewhat better.) Recall that a subset S of complex numbers is called **relatively compact** if its closure is compact (closed and bounded). Such a subset S is relatively compact if and only if every sequence in S has a convergent subsequence. (The subsequence is allowed to converge to a point not in S by definition.)

A subset Φ of $\mathrm{Hol}(U)$ is said to be **uniformly bounded on compact sets in U** if for each compact set K in U there exists a positive number $B(K)$ such that

$$|f(z)| \leqq B(K) \qquad \text{for all} \quad f \in \Phi \quad z \in K.$$

A subset Φ of $\mathrm{Hol}(U)$ is said to be **equicontinuous on a compact set K** if, given ϵ, there exists δ such that if $z, z' \in K$ and $|z - z'| < \delta$, then

$$|f(z) - f(z')| < \epsilon \qquad \text{for all} \quad f \in \Phi.$$

Ascoli's theorem from real analysis states that an equicontinuous family of continuous functions on a compact set is relatively compact. We shall actually reprove it in the context of the next theorem.

Theorem 2.1. *Let $\Phi \subset \mathrm{Hol}(U)$, and assume that Φ is uniformly bounded on compact sets in U. Then Φ is relatively compact.*

Proof. The first step consists in proving that Φ is equicontinuous, on each compact set. We then use a diagonal procedure to find the convergent subsequence.

Let K be compact and contained in U. Let $3r$ be the distance from K to the complement of U. Let $z, z' \in K$ and let C be the circle centered at z' of radius $2r$. Suppose that $|z - z'| < r$. We have

$$\frac{1}{\zeta - z} - \frac{1}{\zeta - z'} = \frac{z - z'}{(\zeta - z)(\zeta - z')},$$

whence by Cauchy's formula,

$$f(z) - f(z') = \frac{z - z'}{2\pi i} \int_C \frac{f(\zeta)}{(\zeta - z)(\zeta - z')} \, d\zeta.$$

Therefore

$$|f(z) - f(z')| < |z - z'| \, \|f\|_{K(2r)} \frac{1}{r},$$

where the sup norm of f is taken on the compact set $K(2r)$ consisting of those $z \in U$ such that $\operatorname{dist}(z, K) \leq 2r$. This proves the equicontinuity of the family on K.

Given a sequence $\{f_n\}$ in Φ, we now prove that there exists a subsequence which converges uniformly on K by the standard proof for Ascoli's theorem. Let $\{z_j\}$ be a countable dense subset of K. For each j, the sequence $\{f_n(z_j)\}$ is bounded. There exists a subsequence $\{f_{n,1}\}$ such that

$$\{f_{n,1}(z_1)\}$$

converges. There exists a subsequence $\{f_{n,2}\}$ of $\{f_{n,1}\}$ such that

$$\{f_{n,2}(z_2)\}$$

converges. Proceeding in this manner we get subsequences $f_{n,j}$ such that

$$\{f_{n,j}(z_1)\}, \ldots, \{f_{n,j}(z_j)\}$$

converge. Then the diagonal subsequence $\{f_{n,n}\}$ is such that

$$\{f_{n,n}(z_j)\}$$

converges for each j.

In fact, we now prove that $\{f_{n,n}\}$ converges uniformly on K. Given ϵ, let δ be as in the definition of equicontinuity. Then for some k, the compact set K is contained in the union of discs

$$K \subset D(z_1, \delta) \cup \cdots \cup D(z_k, \delta).$$

Select N such that if $m, n > N$, then

$$|f_{n,n}(z_j) - f_{m,m}(z_j)| < \epsilon \qquad \text{for} \quad j = 1, \ldots, k.$$

Let $z \in K$. Then $z \in D(z_i, \delta)$ for some i, and we get

$$|f_{n,n}(z) - f_{m,m}(z)| \leq |f_{n,n}(z) - f_{n,n}(z_i)|$$
$$+ |f_{n,n}(z_i) - f_{m,m}(z_i)| + |f_{m,m}(z_i) - f_{m,m}(z)|.$$

The first and third term are $< \epsilon$ by the definition of equicontinuity. The middle term is $< \epsilon$ by what was just proved. We have therefore obtained a subsequence of the original sequence which converges uniformly on K.

We now perform another diagonal procedure.

Lemma 2.2. *There exists a sequence of compact sets K_s $(s = 1, 2, \ldots)$ such that K_s is contained in the interior of K_{s+1} and such that the union of all K_s is U.*

Proof. Let $E(s)$ be the closed disc of radius s, let Z be the closure of U, and let

$$K_s = \text{set of points } z \in Z \cap E(s) \text{ such that}$$
$$\text{dist}(z, \text{boundary } Z) \geq 1/s.$$

Then K_s is contained in the interior of K_{s+1}. For instance, K_s is contained in the open set of elements $z \in U \cap D(0, s + 1)$ such that

$$\text{dist}(z, \text{boundary } Z) > \frac{1}{s + 1}.$$

It is clear that the union of all K_s is equal to U. It then follows that any compact set K is contained in some K_s because the union of these open sets covers U, and a finite number of them covers K.

Let $\{f_n\}$ be the original sequence in Φ. There exists a subsequence $\{f_{n,1}\}$ which converges uniformly on K_1, then a subsequence $\{f_{n,2}\}$ which converges uniformly on K_2, and so forth. The diagonal sequence

$$\{f_{n,n}\}$$

converges uniformly on each K_s, whence on every compact set, and the theorem is proved.

EXERCISES XIV §2

Let U be an open set, and let $\{K_s\}$ ($s = 1, 2, \ldots$) be a sequence of compact subsets of U such that K_s is contained in the interior of K_{s+1} for all s, and the union of the sets K_s is U. For f holomorphic on U, define

$$\sigma_s(f) = \min(1, \|f\|_s),$$

where $\|f\|_s$ is the sup norm of f on K_s. Define

$$\sigma(f) = \sum_{s=1}^{\infty} \frac{1}{2^s} \sigma_s(f).$$

1. Prove that σ is a norm on $\mathrm{Hol}(U)$.

2. Prove that a sequence $\{f_n\}$ in $\mathrm{Hol}(U)$ converges uniformly on every compact subset of U if and only if it converges for the norm σ.

3. Prove that $\mathrm{Hol}(U)$ is complete under the norm σ.

4. Prove that the map $f \mapsto f'$ is a continuous map of $\mathrm{Hol}(U)$ into itself, for the norm σ.

5. Show that a subset of $\mathrm{Hol}(U)$ is relatively compact in the sense defined in the text if and only if it is relatively compact with respect to the norm σ in the usual sense, namely its closure is compact.

Using these exercises, you may then combine the fact that Φ is equicontinuous in Theorem 2.1, with the usual statement of Ascoli's theorem, *without reproving the latter ad hoc*, to conclude the proof of Theorem 2.1.

6. Let Φ be the family of all analytic functions

$$f(z) = z + a_2 z^2 + a_3 z^3 + \cdots$$

on the open unit disc, such that $|a_n| \leqq n$ for each n. Show that Φ is relatively compact.

XIV §3. Proof of the Riemann Mapping Theorem

The theorem will be proved by considering an appropriate family of mappings, and maximizing the derivatives at one point. We first make a reduction which makes things technically simpler later. We let D be the unit disc $D(0, 1)$.

*Let U be a simply connected open set $\neq$ **C**, and let $z_0 \in U$.* We consider the family of all functions

$$f : U \to D$$

such that $f(z_0) = 0$ and such that f is injective. We shall prove that this family is not empty, and it is uniformly bounded. We shall then prove that there exists an element in the family for which $|f'(z_0)|$ is maximal in the family, and that this element gives the desired isomorphism of U and D. Since the only automorphisms of D leaving the origin fixed are rotations, we may then rotate such f so that $f'(z_0)$ is real and positive, thus determining f uniquely. We now carry out the program.

Lemma 3.1. *Let U be an open connected set. Let*

$$f : U \to \mathbf{C}$$

be analytic and injective. Then $f'(z) \neq 0$ for all $z \in U$, and f is an analytic isomorphism of U and its image.

Proof. If $f'(z_0) = 0$ for some $z_0 \in U$, then f has a power series expansion

$$f(z) = a_0 + a_m(z - z_0)^m + \cdots, \qquad a_m \neq 0,$$

with $m \geq 2$, and we know from Chapter II, §5 that f can be decomposed locally near z_0 into a local isomorphism and an m-th power. But locally, the m-th power is not injective, and we conclude therefore that $m = 1$. Thus f is a local analytic isomorphism at every point. Since it is injective, and is an open mapping, it is an analytic isomorphism of U with its image. This proves the lemma.

Lemma 3.2. *Let U be a connected open set, and let $\{f_n\}$ be a sequence of injective analytic maps of U into **C** which converges uniformly on every compact subset of U. Then the limit function f is either constant or injective.*

Proof. Suppose that f is not injective, so there exist two points $z_1 \neq z_2$ in U such that

$$f(z_1) = f(z_2) = \alpha.$$

Let $g_n = f_n - f_n(z_1)$. Then $\{g_n\}$ is a sequence which converges uniformly to

$$g = f - f(z_1).$$

Since f_n is assumed injective, it follows that g_n has no zero on U except at z_1. Suppose that g is not identically zero, and hence not identically zero near z_2 since we assumed that U is connected. Then z_2 is an isolated zero of g. Let γ be a small circle centered at z_2. Then $g(z)$ has a lower bound $\neq 0$ on γ (which is compact), and $\{1/g_n\}$ converges uniformly to $1/g$ on γ. By the formula of Chapter VI, §1 we know that

$$0 = \operatorname{ord}_{z_2} g_n = \frac{1}{2\pi i} \int \frac{g_n'(z)}{g_n(z)} \, dz.$$

Taking the limit as $n \to \infty$ we conclude that $\operatorname{ord}_{z_2} g = 0$ also, a contradiction which proves the lemma.

We now come to the main proof. We first make a reduction. We can always find some isomorphism of U with an open subset of the disc. To see this, we use the assumption that there exists some point $\alpha \in \mathbf{C}$ and $\alpha \notin U$. Since U is simply connected, there exists a determination

$$g(z) = \log(z - \alpha)$$

for $z \in U$ which is analytic on U. This function g is injective, for

$$g(z_1) = g(z_2) \Rightarrow e^{g(z_1)} = e^{g(z_2)} \Rightarrow z_1 - \alpha = z_2 - \alpha,$$

whence $z_1 = z_2$. If g takes on some value $g(z_0)$, then

$$g(z) \neq g(z_0) + 2\pi i$$

for all $z \in U$, as one sees again by exponentiating. We claim that there exists a disc around $g(z_0) + 2\pi i$ such that g takes on no value in this disc. Otherwise, there is a sequence $w_n \in U$ such that $g(w_n)$ approaches $g(z_0) + 2\pi i$, and exponentiating shows that w_n approaches z_0, so $g(w_n)$ approaches $g(z_0)$, a contradiction. Hence the function

$$\frac{1}{g(z) - g(z_0) - 2\pi i}$$

is bounded on U, and is analytic injective. By a translation and multiplication by a small positive real number, we may then obtain a function f such that $f(z_0) = 0$ and $|f(z)| < 1$ for all $z \in U$. This function f is injective, and so an isomorphism of U onto an open subset of the disc.

To prove the Riemann mapping theorem, we may now assume without loss of generality that U is an open subset of D and contains the origin. Let Φ be the family of all injective analytic maps

$$f : U \to D$$

such that $f(0) = 0$. This family is not empty, as it contains the identity. Furthermore, the absolute values $|f'(0)|$ for $f \in \Phi$ are bounded. This is obvious from Cauchy's formula

$$2\pi i f'(0) = \int \frac{f(\zeta)}{\zeta^2} \, d\zeta$$

and the uniform boundedness of the functions f, on some small closed disc centered at the origin. The integral is taken along the circle bounding the disc.

Let λ be the least upper bound of $|f'(0)|$ for $f \in \Phi$. Let $\{f_n\}$ be a sequence in Φ such that $|f'_n(0)| \to \lambda$. Picking a subsequence if necessary, Theorem 2.1 implies that we can find a limit function f such that

$$|f'(0)| = \lambda,$$

and $|f(z)| \leq 1$ for all $z \in U$. Lemma 3.2 tells us that f is injective, and the maximum modulus principle tells us that in fact,

$$|f(z)| < 1 \qquad \text{for all} \quad z \in U.$$

Therefore $f \in \Phi$, and $|f'(0)|$ is maximal in the family Φ. The next lemma concludes the proof of the Riemann mapping theorem.

Lemma 3.3. *Let $f \in \Phi$ be such that $|f'(0)| \geq |g'(0)|$ for all $g \in \Phi$. Then f is an analytic isomorphism of U with the disc.*

Proof. All we have to prove is that f is surjective. Suppose not. Let $\alpha \in D$ be outside the image of f. Let T be an automorphism of the disc such that $T(\alpha) = 0$ (for instance, $T(z) = (\alpha - z)/(1 - \bar{\alpha}z)$, but the particular shape is irrelevant here). Then $T \circ f$ is an isomorphism of U onto an open subset of D which does not contain 0, and is simply connected since U is simply connected. We can therefore define a square root function on $T(f(U))$, for instance we can define

$$\sqrt{T(f(z))} = \exp(\tfrac{1}{2} \log T(f(z))) \qquad \text{for} \quad z \in U.$$

Note that the map

$$z \mapsto \sqrt{T(f(z))} \qquad \text{for} \quad z \in U$$

is injective, because if z_1, z_2 are two elements of U at which the map takes the same value, then $Tf(z_1) = Tf(z_2)$, whence $f(z_1) = f(z_2)$ because T is injective, and $z_1 = z_2$ because f is injective. Furthermore $T(f(U))$ is contained in D, and does not contain 0.

Let R be an automorphism of D which sends $T(f(0))$ to 0. Then

$$g: z \mapsto R\left(\sqrt{Tf(z)}\right)$$

is an injective map of U into D such that $g(0) = 0$, so $g \in \Phi$. It suffices to prove that $|g'(0)| > |f'(0)|$ to finish the proof. But if we let S be the square function, and we let

$$\varphi(w) = T^{-1}(S(R^{-1}(w))),$$

then

$$f(z) = \varphi(g(z)).$$

Furthermore, $\varphi: D \to D$ is a map of D into itself, such that $\varphi(0) = 0$, and φ is not injective because of the square function S. By Theorem 1.2 of Chapter VII (the complement to the Schwarz lemma), we know that $|\varphi'(0)| < 1$. But

$$f'(0) = \varphi'(0)g'(0),$$

so $|g'(0)| > |f'(0)|$, contradicting our assumption that $|f'(0)|$ is maximal. This concludes the proof.

XIV §4. Behavior at the Boundary

We investigate the extent to which an isomorphism

$$f: U \to D$$

of U with the disc can be extended by continuity to the boundary of U. There is a standard picture which shows the type of difficulty (impossibility) which can happen if the boundary is too complicated. In Fig. 5,

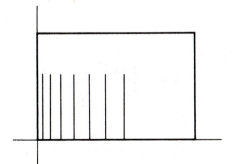

Figure 5

the open set U consists of the interior of the rectangle, from which vertical segments as shown are deleted. These segments have their base point at $1/n$ for $n = 1, 2, \ldots$. It can be shown that for such an open set U there is no way to extend the Riemann mapping function by continuity to the origin, which in some sense is "inaccessible".

On the other hand, we shall now prove that in all cases of interest, the mapping can so be extended.

Throughout this section, the word **curve** will mean **continuous** curve. No further smoothness is needed, and in fact it is useful to have the flexibility of continuity to work with.

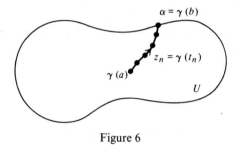

Figure 6

Let U be an open set, and α a boundary point. We say that α is **accessible** if given a sequence of points $z_n \in U$ such that $\lim z_n = \alpha$, there exists a continuous curve

$$\gamma: [a, b] \to \mathbf{C}$$

such that $\gamma(b) = \alpha$, $\gamma(t) \in U$ for all $t \neq b$, and there exist $t_n \in [a, b]$ such that $\gamma(t_n) = z_n$ (in other words, the curve passes through the given sequence), and $a < t_1 < t_2 < \cdots$, $\lim t_n = b$. Note that the curve lies entirely in U except for its end point.

Theorem 4.1. *Let U be simply connected and bounded, and let*

$$f: U \to D$$

be an isomorphism with the disc. If α is an accessible boundary point of U, then

$$\lim_{z \to \alpha} f(z)$$

exists for $z \in U$.

Proof. Suppose not. Then there exists a sequence $\{z_n\}$ in U tending to α, but $\{f(z_n)\}$ has no limit. We find a curve γ as in the definition of accessibility.

Lemma 4.2. $\lim\limits_{t \to b} |f(\gamma(t))| = 1.$

Proof. Suppose not. Given ϵ there exists a sequence of increasing numbers s_n such that $|f(\gamma(s_n))| \leq 1 - \epsilon$, and taking a subsequence if necessary, we may assume $f(\gamma(s_n))$ converges to some point w with $|w| \leq 1 - \epsilon$. Let

$$g: D \to U$$

be the inverse function to f. Then

$$\gamma(s_n) = g(f(\gamma(s_n))) = \gamma(s_n)$$

and $\gamma(s_n) \to g(w)$, so $\gamma(s_n)$ cannot tend to the boundary of U, a contradiction which proves the lemma.

Since $\{f(z_n)\}$ has no limit, and since the closed disc D^c is compact, there exist two subsequences of $\{z_n\}$, which we will denote by $\{z'_n\}$ and $\{z''_n\}$, such that $f(z'_n)$ tends to a point w' on the unit circle (by the lemma), and $f(z''_n)$ tends to a point $w'' \neq w'$ on the unit circle. The continuous curve γ then restricts to a curve γ_n joining z'_n and z''_n, and γ_n comes close to α as n becomes large. The image $f(\gamma_n)$ is a curve in D, joining $f(z'_n) = w'_n$ to $f(z''_n) = w''_n$, and $f(\gamma_n)$ tends to the unit circle uniformly by the lemma. The picture is as on Fig. 7.

Let us draw rays from the origin to points on the circle close to w', and also two rays from the origin to points on the circle close to w'', as shown on the figure. We take n sufficiently large. Then for infinitely many n, the curves $f(\gamma_n)$ will lie in the same smaller sector. Passing to a subsequence if necessary, we may assume that for all n, the curves $f(\gamma_n)$ lie in the same sector.

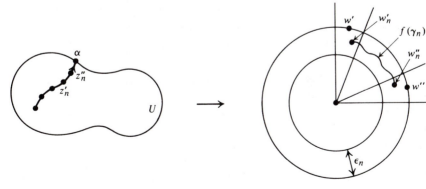

Figure 7

As before, we let $g: D \to U$ be the inverse function to f, and we let

$$h = g - \alpha,$$

so $h(w) = g(w) - \alpha$ for $|w| < 1$. Then $h(f(\gamma_n)) \to 0$ as $n \to \infty$. We wish to apply the maximum modulus, but we have to make an auxiliary construction of another function which takes on small values on a curve surrounding the whole circle of radius $1 - \epsilon_n$. This is done as follows. We state a general lemma, due to Lindelöf and Koebe, according to Bieberbach.

Lemma 4.3. *Let h be analytic on the unit disc D, and bounded. Let w', w'' be two distinct points on the circle. Let $\{w'_n\}$ and $\{w''_n\}$ be sequences in the unit disc D converging to w' and w'', respectively, and let ψ_n be a curve joining w'_n with w''_n such that ψ_n lies in the ring*

$$1 - \epsilon_n < |w| < 1,$$

and $\epsilon_n \to 0$ as $n \to \infty$. Assume that $r_n > 0$ is a sequence tending to 0 such that

$$|h(w)| < r_n \quad \text{for} \quad w \text{ on } \psi_n.$$

Then h is identically 0.

Proof. Dividing $h(w)$ by some power of w if necessary, we may assume without loss of generality that $h(0) \neq 0$. Without loss of generality, after a rotation, we may assume that

$$w'' = \overline{w'},$$

so w' and w'' are symmetric about the horizontal axis. We pick a large integer M and let L' be the ray having angle $2\pi/2M$ with the x-axis. Let L'' be the ray having angle $-2\pi/2M$ with the axis, as shown on Fig. 8.

We let ψ_n be defined on an interval $[a_n, b_n]$. Let u_n be the largest value of the parameter such that $\psi_n(u_n)$ is on L', and let v_n be the smallest value of the parameter $> u_n$ such that $\psi_n(v_n)$ lies on the x-axis. Then ψ_n restricted to $[u_n, v_n]$ is a curve inside the sector lying between the x-axis and L', and connecting the point $\psi_n(u_n)$ with the point $\psi_n(v_n)$. If we reflect this curve across the x-axis, then we obtain a curve which we denote by $\overline{\psi}_n$, joining $\psi_n(v_n)$ with $\overline{\psi_n(u_n)}$. We let σ_n be the join of these two curves, so that σ_n is symmetric about the x-axis, and joins $\psi_n(u_n)$ with $\overline{\psi_n(u_n)}$,

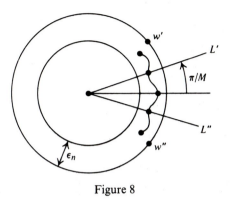

Figure 8

passing through $\psi_n(v_n)$, as shown on Fig. 9. Then the beginning and end points of σ_n lie at the same distance from the origin. (This is what we wanted to achieve to make the next step valid.) Let T be rotation by the angle $2\pi/M$. If we rotate σ_n by T iterated M times, i.e. take

$$\sigma_n, T\sigma_n, \ldots, T^{M-1}\sigma_n,$$

then we obtain a closed curve, lying inside the annulus $1 - \epsilon_n < |w| < 1$. Finally, define the function

$$h^*(w) = h(w)\overline{h(\overline{w})},$$

and the function

$$G(w) = h^*(w)h^*(Tw) \cdots h^*(T^{M-1}w).$$

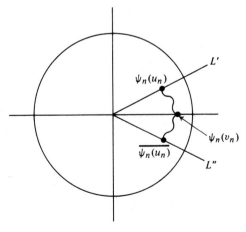

Figure 9

Let B be a bound for $|h(w)|$, $w \in D$. Each factor in the definition of G is bounded by B^2. For any w on the above closed curve, some rotation $T^k w$ lies on σ_n, and then we have

$$|h^*(w)| \leq r_n B.$$

Therefore G is bounded on the closed curve by

$$|G(w)| \leq r_n B^{2M-1}.$$

For each ray L from the origin, let w_L be the point of L closest to the origin, and lying on the closed curve. Let W be the union of all segments $[0, w_L]$ open on the right, for all rays L. Then W is open, and the boundary of W consists of points of the closed curve. By the maximum modulus principle, it follows that

$$|G(0)| \leq \max |G(w)|,$$

where the max is taken for w on the closed curve. Letting n tend to infinity, we see that $G(0) = 0$, whence $h(0) = 0$, a contradiction which proves the lemma, and therefore also the theorem.

Theorem 4.4. *Let U be bounded. Let $f : U \to D$ be an isomorphism with the disc, and let α_1, α_2 be two distinct boundary points of U which are accessible. Suppose f extended to α_1 and α_2 by continuity. Then*

$$f(\alpha_1) \neq f(\alpha_2).$$

Proof. We suppose $f(\alpha_1) = f(\alpha_2)$. After multiplying f by a suitable constant, we may assume $f(\alpha_i) = -1$. Let

$$g : D \to U$$

again be the inverse function of f. Let γ_1, γ_2 be the curves defined on an interval $[a, b]$ such that their end points are α_1, α_2, respectively, and $\gamma_i(t) \in U$ for $i = 1, 2$ and $t \in [a, b]$, $t \neq b$. There exists a number c with $a < c < b$ such that

$$|\gamma_1(t) - \gamma_2(t)| > \tfrac{1}{2}|\alpha_1 - \alpha_2|, \qquad \text{if} \quad c < t < b,$$

and there exists δ such that

$$f(\gamma_1([a, c])) \qquad \text{and} \qquad f(\gamma_2([a, c]))$$

do not intersect the disc $D(-1, \delta)$ as shown on Fig. 10. Let

$$A(\delta) = D \cap D(-1, \delta).$$

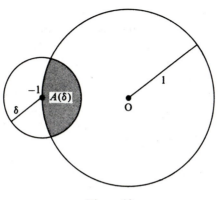

Figure 10

Then $A(\delta)$ is described in polar coordinates by

$$0 \leq r \leq \delta \quad \text{and} \quad -\varphi(r) \leq \theta \leq \varphi(r)$$

with an appropriate function $\varphi(r)$. Note that $\varphi(r) < \pi/2$. We have:

$$\text{Area } g(A(\delta)) = \iint_{A(\delta)} |g'(z)|^2 \, dy \, dx$$

$$= \int_0^\delta \int_{-\varphi(r)}^{\varphi(r)} |g'(-1 + re^{i\theta})|^2 r \, d\theta \, dr.$$

For each $r < \delta$ let w_1, w_2 be on $f(\gamma_1)$, $f(\gamma_2)$, respectively, such that

$$|w_i + 1| = r, \quad i = 1, 2,$$

and

$$|g(w_1) - g(w_2)| > \tfrac{1}{2}|\alpha_1 - \alpha_2|.$$

Then

$$g(w_1) - g(w_2) = \int_{w_1}^{w_2} g',$$

where the integral is taken over the circular arc from w_1 to w_2 in D, with center -1. For $0 < r < \delta$ we get

$$\tfrac{1}{2}|\alpha_1 - \alpha_2| < \int_{-\varphi(r)}^{\varphi(r)} |g'(-1 + re^{i\theta})| r \, d\theta,$$

whence by the Schwarz inequality, we find

$$\frac{|\alpha_1 - \alpha_2|^2}{4\pi r} \leq r \int_{-\varphi(r)}^{\varphi(r)} |g'(-1 + re^{i\theta})|^2 \, d\theta.$$

We integrate both sides with respect to r from 0 to δ. The right-hand side is bounded, and the left-hand side is infinite unless $\alpha_1 = \alpha_2$. This proves the theorem.

The technique for the above proof is classical. It can also be used to prove the continuity of the mapping function at the boundary. Cf. for instance Hurwitz–Courant, Part III, Chapter 6, §4.

APPENDIX

Cauchy's Formula for C^∞ Functions

Let D be an open disc in the complex numbers, and let D^c be the closed disc, so the boundary of D^c is a circle. Cauchy's formula gives us the value as an integral over the circle C:

$$f(z_0) = \frac{1}{2\pi i} \int_C \frac{f(z)\,dz}{z - z_0}$$

if f is holomorphic on D^c, that is on some open set containing the closed disc. But what happens if f is not holomorphic but merely smooth, say its real and imaginary parts are infinitely differentiable in the real sense? There is also a formula, which unfortunately is not usually taught in basic courses, although it gives a beautiful application of several notions which arise in both real and complex analysis, and advanced calculus. We shall give this theorem here, together with an application, which occurs in the theory of partial differential equations.

Let us write $z = x + iy$. We introduce two new derivatives. Let

$$f(z) = f_1(x, y) + if_2(x, y),$$

where $f_1 = \operatorname{Re} f$ and $f_2 = \operatorname{Im} f$ are the real and imaginary parts of f respectively. We say that f is C^∞ if f_1, f_2 are infinitely differentiable in the naive sense of functions of two real variables x and y. In other words, all partial derivatives of all orders exist and are continuous. We write $f \in C^\infty(D^c)$ to mean that f is C^∞ on some open set containing D^c.

For such f we define

$$\frac{\partial f}{\partial z} = \frac{1}{2}\left(\frac{\partial f}{\partial x} - i\frac{\partial f}{\partial y}\right) \quad \text{and} \quad \frac{\partial f}{\partial \bar{z}} = \frac{1}{2}\left(\frac{\partial f}{\partial x} + i\frac{\partial f}{\partial y}\right).$$

Symbolically, we put

$$\frac{\partial}{\partial z} = \frac{1}{2}\left(\frac{\partial}{\partial x} - i\frac{\partial}{\partial y}\right) \quad \text{and} \quad \frac{\partial}{\partial \bar{z}} = \frac{1}{2}\left(\frac{\partial}{\partial x} + i\frac{\partial}{\partial y}\right).$$

The **Cauchy–Riemann equations** can be formulated neatly by saying that f is holomorphic if and only if

$$\frac{\partial f}{\partial \bar{z}} = 0.$$

See Chapter VIII, §1.

We shall need the Stokes–Green formula for a simple type of region. In advanced calculus, one integrates expressions

$$\int_C P\,dx + Q\,dy,$$

where P, Q are C^∞ functions, and C is some curve. The Stokes–Green theorem relates such integrals over a boundary to a double integral

$$\iint_B \left(\frac{\partial Q}{\partial x} - \frac{\partial P}{\partial y}\right) dx\,dy$$

taken over a region B which is bounded by the curve C. The precise statement is this.

Stokes–Green Formula. *Let B be a region of the plane, bounded by a finite number of curves, oriented so that the region lies to the left of each curve. Let γ be the boundary, so oriented. Let P, Q have continuous first partial derivatives on B and its boundary. Then*

$$\int_\gamma P\,dx + Q\,dy = \iint_B \left(\frac{\partial Q}{\partial x} - \frac{\partial P}{\partial y}\right) dx\,dy.$$

It is useful to express the Stokes–Green formula in terms of the derivatives $\partial/\partial z$ and $\partial/\partial \bar{z}$. Writing

$$dz = dx + i\,dy \quad \text{and} \quad d\bar{z} = dx - i\,dy,$$

we can solve for dx and dy in terms of dz and $d\bar{z}$, to give

$$dx = \tfrac{1}{2}(dz + d\bar{z}) \quad \text{and} \quad dy = \frac{1}{2i}(dz - d\bar{z}).$$

Then

$$P\,dx + Q\,dy = g\,dz + h\,d\bar{z},$$

where g, h are suitable functions. Let us write symbolically

$$dz \wedge d\bar{z} = -2i\,dx\,dy.$$

Then by substitution, we find the following version of the **Stokes–Green Formula**:

$$\int_\gamma g\,dz + h\,d\bar{z} = \iint_B \left(\frac{\partial h}{\partial z} - \frac{\partial g}{\partial \bar{z}}\right) dz \wedge d\bar{z}.$$

Remark. Directly from the definition of $\partial/\partial z$ and $\partial/\partial \bar{z}$ one verifies that the usual expression for df is given by

$$\frac{\partial f}{\partial z}\,dz + \frac{\partial f}{\partial \bar{z}}\,d\bar{z} = \frac{\partial f}{\partial x}\,dx + \frac{\partial f}{\partial y}\,dy.$$

Consider the special case where $B = B(a)$ is the region obtained from the disc D^c by deleting a small disc of radius a centered at a point z_0, as shown on the figure.

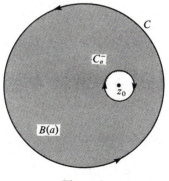

Figure 1

Then the boundary consists of two circles C and C_a^-, oriented as shown so that the region lies to the left of each curve. We have written C_a^- to indicate the circle with clockwise orientation, so that the region $B(a)$ lies to the left of C_a^-. As before, C is the circle around D, oriented counterclockwise. Then the boundary of $B(a)$ can be written

$$\gamma = C + C_a^-.$$

We shall deal with integrals

$$\iint_D \varphi(z) \frac{dz \wedge d\bar{z}}{z - z_0},$$

where $\varphi(z)$ is a smooth function, and where z_0 is some point in the interior of the disc. Such an integral is an improper integral, and is supposed to be interpreted as a limit

$$\lim_{a \to 0} \iint_{B(a)} \varphi(z) \frac{dz \wedge d\bar{z}}{z - z_0},$$

where $B(a)$ is the complement of a disc of radius a centered at z_0. The limit exists, as one sees by using polar coordinates. Letting $z = z_0 + re^{i\theta}$ with polar coordinates around the fixed point z_0, we have

$$dx\, dy = r\, dr\, d\theta$$

and $z - z_0 = re^{i\theta}$, so r cancels and we see that the limit exists, since the integral becomes simply

$$\iint_D \varphi(z)\, dr\, d\theta.$$

The region $B(a)$ is precisely of the type where we apply the Stokes–Green formula.

Cauchy's Theorem for C^∞ functions. *Let $f \in C^\infty(D^c)$ and let z_0 be a point in the interior D. Let C be the circle around D. Then*

$$f(z_0) = \frac{1}{2\pi i} \int_C \frac{f(z)\, dz}{z - z_0} + \frac{1}{2\pi i} \iint_D \frac{\partial f}{\partial \bar{z}} \frac{dz \wedge d\bar{z}}{z - z_0}.$$

Proof. Let a be a small positive number, and let $B(a)$ be the region obtained by deleting from D a disc of radius a centered at z_0. Then $\partial f/\partial \bar{z}$ is C^∞ on $B(a)$ and we can apply the Stokes–Green formula to

$$\frac{f(z)\, dz}{z - z_0} = g(z)\, dz$$

over this region. Note that this expression has no term with $d\bar{z}$. Furthermore

$$\frac{\partial}{\partial \bar{z}} \left(\frac{1}{z - z_0} \right) = 0 \quad \text{and} \quad \frac{\partial g}{\partial \bar{z}} = \frac{\partial f}{\partial \bar{z}} \frac{1}{z - z_0}$$

because $1/(z - z_0)$ is holomorphic in this region. Then by Stokes–Green we find

$$\int_{C_a^-} \frac{f(z)\, dz}{z - z_0} + \int_C \frac{f(z)\, dz}{z - z_0} = - \iint_{B(a)} \frac{\partial f}{\partial \bar{z}} \frac{dz \wedge d\bar{z}}{z - z_0}.$$

The limit of the integral on the right-hand side as a approaches 0 is the double integral (with a minus sign) which occurs in Cauchy's formula. We now determine the limit of the curve integral over C_a^- on the left-hand side. We parametrize C_a (counterclockwise orientation) by

$$z = z_0 + ae^{i\theta}, \qquad 0 \le \theta \le 2\pi.$$

Then $dz = aie^{i\theta}\, dz$, so

$$\int_{C_a^-} \frac{f(z)\, dz}{z - z_0} = - \int_{C_a} \frac{f(z)\, dz}{z - z_0} = - \int_0^{2\pi} f(z_0 + ae^{i\theta})i\, d\theta.$$

Since f is continuous at z_0, we can write

$$f(z_0 + ae^{i\theta}) = f(z_0) + h(a, \theta)$$

where $h(a, \theta)$ is a function such that

$$\lim_{a \to 0} h(a, \theta) = 0$$

uniformly in θ. Therefore

$$\lim_{a \to 0} \int_{C_a^-} \frac{f(z)\, dz}{z - z_0} = -2\pi i f(z_0) - \lim_{a \to 0} \int_0^{2\pi} h(a, \theta)i\, d\theta$$

$$= -2\pi i f(z_0).$$

Cauchy's formula now follows at once.

Remark 1. Suppose that f is holomorphic on D. Then

$$\frac{\partial f}{\partial \bar{z}} = 0,$$

and so the double integral disappears from the general formula to give the Cauchy formula as we encountered it previously.

Remark 2.   Consider the special case when the function f is 0 on the boundary of the disc.   Then the integral over the circle C is equal to 0, and we obtain the formula

$$f(z_0) = \frac{1}{2\pi i} \iint_D \frac{\partial f}{\partial \bar{z}} \frac{dz \wedge d\bar{z}}{z - z_0}.$$

This allows us to recover the values of the function from its derivative $\partial f / \partial \bar{z}$.   Conversely, one has the following result.

Theorem.   Let $g \in C^\infty(D^c)$ be a C^∞ function on the closed disc.   Then the function

$$f(w) = \frac{1}{2\pi i} \iint_D \frac{g(z)}{z - w} dz \wedge d\bar{z}$$

is defined and C^∞ on D, and satisfies

$$\frac{\partial f}{\partial \bar{w}} = g(w) \qquad for \quad w \in D.$$

The proof is essentially a corollary of Cauchy's theorem if one has the appropriate technique for differentiating under the integral sign.   However, we have now reached the boundary of this course, and we omit the proof.

Index

Graduate Texts in Mathematics

continued from page ii